Monopulse Measurement with Active Electronically Scanned Arrays (AESAs)

Roger A. Dana

Monopulse Measurement with Active Electronically Scanned Arrays (AESAs)

 Springer

Roger A. Dana
RAD Concepts LLC
Mount Vernon, IA, USA

ISBN 978-3-030-91910-8   ISBN 978-3-030-91908-5   (eBook)
https://doi.org/10.1007/978-3-030-91908-5

This Springer imprint is published by the registered company Springer Nature Switzerland AG
The registered company address is: Gewerbestrasse 11, 6330 Cham, Switzerland

*This book would not have been written
without the encouragement and love I
received from Armella M. Dana (1948–2017),
my wife of nearly 45 years, who made me a
better person and who still inspires me to
excellence, something she always strove for in
her art and music.*

Preface

This monograph sits atop 40 plus years in the aerospace industry, after an academic career that culminated in a Ph.D. in astrophysics. Some of my first work at Hughes Aircraft Company involved evaluating the performance of detection algorithms for air defense radars, work done after I had developed some understanding[1] of radar detection in noise and clutter environments. Most of that work was done with a piece of paper and a pencil (and a big eraser as one rarely gets it right the first time). I learned that under many circumstances, one could develop considerable insight into the problem by just writing it out to ensure capture of the essential elements and investigating what could be learned from "analytic, closed form calculations." Later, at Mission Research Corporation (MRC) as the problems became more complex and usually nonlinear in some fashion, I learned how to put together efficient simulations that shed light on how the system actually worked with multiple things going on at once. I still strove to develop analytic solutions to check the simulation results in those simplified cases where both could be done in a similar manner. I found that my "customers," whether in-house or in Government, appreciated some assurance that the simulations were verified before they were used to make conclusions about the problem at hand. I finished my career as an employee at Collins Aerospace Advanced Technology Center (ATC), where I worked on a variety of problems ranging from communications, to satellite and inertial navigation, to weather radar and conventional radar, to tracking and location with optical systems, to Kalman filtering, to understanding Active Electronically Scanned Arrays (AESAs). Few of these lent themselves to pure analysis, but all benefited by first enumerating the applicable "first principles" before developing solutions.

By first principles, I mean the basic ideas, usually expressed mathematically, that govern the problem. For example, the predecessor to this book (Dana 2019) starts

---

[1] Understanding gained, in part, from a short course on radar principles taught by Merrill I. Skolnik that I took in my first few months at Hughes, and from which I received a copy of Skolnik (1962), an old friend over the years.

with Maxwell's equations to illustrate that plane waves that are used to compute AESA gain are indeed solutions. But for monopulse, the first principles are the expressions that are used to measure the Angle of Arrival (AoA) of the signal, whether phase- or amplitude-comparison monopulse. In other problems, it is a starting point that will be recognized by informed readers as the truth from which the problem develops.

It was at ATC that I began to ponder more often what are the essential ingredients to the problems that came my way. Sometimes that is so obvious that it takes no time. But then there are those problems that could involve many different aspects, but that in essence could be more compactly analyzed and still provide results that "answer the mail." It is clear that one cannot simulate the universe and get an answer to one's customer within the allotted level of effort and schedule. Often it is less clear exactly what parts of the universe contribute in a significant way to the problem, which of those parts actually can be modeled with any fidelity, and which parts are necessary to model in the ideal case. For example, we know that AESAs in real environments encounter vibration and flexing, unwanted temporary coatings of water, ice, or dust, and possibly huge temperature fluctuations in going from a hot tarmac to the environment at 35,000 ft altitude. Furthermore, in the environment there is possible multipath and jamming interference, and atmospheric attenuation. Other effects include element mutual coupling, impedance modulation as a function of scan angle, and surface wave initiation with its possibility of scan blindness [see, e.g., Mailloux (2005)]. But these effects are not initially important to verifying whether or not a monopulse algorithm implemented in an AESA will meet its requirements with some comfortable margin to allow the real world to happen. Later, the systems engineer may need to visit the real-world problems, but will be better prepared to deal with them in front of a program manager[2] after quantifying "ideal performance" and its margin relative to system requirements.

So, this work is both a treatise on two forms of monopulse that can be implemented in AESAs, and an example of how to evaluate the performance of an electronic system via both simulations and "back of the envelope" analytic calculations. The goal is to show how "first principles" of monopulse algorithms and AESA signals and noise are combined to simulate the performance of these systems and then to understand key limitations of that performance.

Mount Vernon, IA, USA   Roger A. Dana, Ph.D.

---

[2]This is my term of endearment for an entity that demands that increases in complexity must be carefully justified, balancing benefit with cost. That entity may be a real person or one's own internal engineering judgment at work.

Acknowledgments

I am grateful for the opportunities to learn and contribute in significant ways that I encountered during my time at Hughes Aircraft Company Ground Systems Group (3 years), Mission Research Corporation (24 years), and Collins Aerospace Advanced Technology Center (15 years). Most of my supervisors over the years were supportive of my approach to problems and to my spending time to look into additional technical questions that arose. This freedom, gained because I produced "good work" and was able to explain my results in a straightforward manner, allowed me more opportunities to work on problems at the intersection of math, physics, and electronic systems. Many individuals, too many to mention save a few, contributed in positive ways to my career and understanding.

Although I had a deep understanding of the concepts behind electronic systems from my earlier career, it was at ATC that I really began to appreciate the art of building radios. A few artisan engineers stand out because they "knew their stuff" and because they were willing to listen to theoretical explanations of why things work. It is in the reconciliation of the theory with the practical engineering insight that full understanding occurred. Many thanks to Gary Lehtola, Dave Gribble, and Don Hovda.

More recently at ATC, these engineering practitioners included Dr. Lee Paulsen, Anders Walker, and James West who supported my work technically and financially, and Aimee Matland-Dixon, Dr. Jiwon Moran, Dr. Matilda Livadaru, and Jeremiah Wolf, in particular, who continually asked probing questions and demanded rigorous answers.

Dr. Patrick Y. C. Hwang deserves special mention, as he taught me much of what I know about Extended Kalman Filters, first through his book on Kalman filtering (Brown and Hwang 1997) and then as we worked together on various projects. These gadgets can be incredibly frustrating to get to work, but once working they can be delightfully rich in behavior.

Timothy Rand, a friend and former colleague at ATC, carefully read a draft of this book, provided many helpful comments, and pointed out a number of typos and notational inconsistencies. His efforts added greatly to this work.

It is with great pleasure that I recall the many discussions in my office at ATC where one of more of these folks would drop by, step to the whiteboard, and begin an inquiry with a sketch or an equation . . .

Contents

Chapter 1
Introduction

Abstract A brief, not exhaustive, discussion on the uses of monopulse in electronic systems, including one-way communications and two-way radars, is presented. Then follows a high-level block diagram of the hardware in an AESA and its controlling Digital Signal Processor, illustrating some of the necessary components of the hardware to support monopulse. Finally, a short, not exhaustive, list of assumed background knowledge is given.

Keywords Active electronically scanned array (AESA) · Angle of arrival (AoA) · Phase comparison monopulse · Amplitude comparison monopulse · Sequential lobing · Signal-to-noise ratio (SNR) · Additive White Gaussian Noise (AWGN)

1.1 Role of Monopulse in Electronic Systems

Monopulse is a technique that is used to measure the Angle-of-Arrival (AoA) of a signal to a precision better (often much better) than the beamwidth of the receiving antenna. The name refers to the fact that the AoA can be measured on a per pulse basis in a radar that is faster than older techniques developed for noncoherent systems. For example, sequential lobing is performed by pointing the antenna at two positions on either side of the radar target and comparing the relative amplitudes of the signal to estimate the AoA. Another technique for scanning radars was to monitor the amplitude of the return as the antenna beam swept past the target, and then pick the pointing angle of the maximum return as the angular location of the target. Both of these techniques require multiple pulses on target to make the AoA measurement. If the two "sequential lobing" measurements are made simultaneously, the technique is called amplitude-comparison monopulse. When two simultaneous phase measurements are made at two separate subarrays to measure the AoA, the technique is referred to as phase-comparison monopulse.

Tracking radar is one application where accurate AoA measurements can be imperative. In the range dimension, pulse compression techniques can be used to obtain range measurement accuracy δR that is much smaller than the transmitted pulsewidth. But in the cross-range dimensions, the target position uncertainty is usually dominated by the range times angle accuracy $R \times \delta\theta$ contribution that can be in two orthogonal directions. In the general case where the target's trajectory is not purely parallel or tangential to the line of sight, the target tracking position error is a combination of these two or three errors, and one does not want $R \times \delta\theta$ to be much larger than δR to get an accurate three-dimensional tracking position.

Of course, monopulse is not limited to radars. One also can measure the AoA of a transmitter in a communication link or an electronic warfare system. However, in the latter case some knowledge of the transmitted waveform may be necessary in order to achieve the necessary signal-to-noise ratio for an accurate measurement (i.e., better than the beamwidth).

In this monograph, we concentrate on phase-comparison monopulse because it is a natural extension of the processes that already occur in an AESA. However, we compare its performance to amplitude-comparison monopulse to show that the former does indeed have better performance, all other things being equal. In both, the amplitude or phase of two adjacent subarrays is compared to measure the AoA. So, in a digital signal processing implementation of an AESA, it would seem (incorrectly) that one could actually do both with the same set of subarray voltages. We will see that the array pointing is different between the two cases, and the subarray outputs are not compatible for both techniques.

We will see also that the monopulse error varies with the square root of the signal-to-noise ratio (SNR). This unfortunate truth is surmountable, however, if one looks for viable options to increase the SNR of the received signals upon which monopulse is performed. To that end, we review the derivations of SNR for one-way and monostatic two-way links in the next chapter.

1.2 Idealized Hardware Implementation

Before discussing subarray-based monopulse implementation, we first need to discuss the idea of having a radio behind each element of an array. To cut off hysterical laughter from program manager types, this is recognized as possible but not practical. Such a design would allow many different simultaneous array applications such as multiple beams, monopulse measurements of multiple targets, simultaneous multi-functions; the list goes on and on. But radios are expensive, and 1024 of them for our example 32×32 element array will add a large factor to the cost of the systems. Perhaps more importantly, elements are spaced at a maximum of one-half the wavelength ($\lambda/2$) apart, and as technology and available bandwidth evolves, we want to operate at higher and higher frequencies in the Ka band, for example, with wavelengths in the millimeter range. That just does not leave enough real estate behind each element to install sufficient electronics for a radio. Indeed, the current challenge for AESA designers is to get control lines into and signals into and out of such small areas and to keep the electronics there from overheating (e.g., a Ka-band power amplifier may have an efficiency of 25 percent at best, meaning that 75 percent of its input power goes into heat, not radiated RF signal).

An idealized hardware implementation of phase- or amplitude-comparison monopulse using four subarrays is shown in Fig. 1.1. The dashed blue line indicates

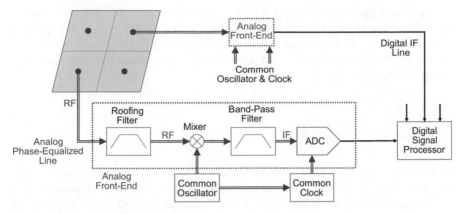

Fig. 1.1 Idealized hardware implementation of phase- or amplitude-comparison monopulse

the analog front end of each of the subarrays with RF inputs indicated by double blue lines. After digitization in an Analog-to-Digital Converter (ADC), the rest of the monopulse processing takes place in a Digital Signal Processor (DSP).

Within the front end, there is a "roofing filter" that attenuates nearby RF signals and sets the initial noise figure of the system. Then the isolated RF is mixed down to an intermediate frequency (IF) signal using a reference tone common to all four front ends. As the mixing produces both sum and difference frequency signals, a band-pass filter is used to eliminate the high frequency content. Finally, the analog IF signal is injected into the Analog-to-Digital Converter (ADC), where the sampling time is triggered by the common clock. The output is a digital IF signal, indicated by solid black lines, four of which are input to the DSP. Of course, there may need to be amplifiers in the analog portion to boost the received signals to the input level required by the ADCs, as each of the proceeding components has an insertion loss and received signals are generally weak.

It is in the DSP that the signals are finally down-converted to baseband forming the in-phase and quadrature phase voltage components that we will represent mathematically as the real and imaginary parts of complex voltages. There is an important distinction that must be recognized when adding together voltages in the DSP versus in analog circuitry. In the former, one is just adding digital representation of voltages; in the latter one has to be careful that energy is conserved in the mathematical representation of the summing process.

For phase-comparison monopulse, a key notation in the figure is "Phase Equalized Line." This means that the phase length of the lines coming from the four subarrays to their analog front ends and the phase paths through those electronics must be the same or nearly so (to within a small fraction of a wavelength). Then the phase differences of the subarray voltages will be due to AoA, not to the electronics. This applies to the mixing tones and the clock triggers also. Once the signals are simultaneously digitized, the DSP can operate on them without having to do phase corrections. Of course, if the phase path differences of the four outputs are different in a deterministic manner, calibration can be used to correct the voltage phases.

Preserving the phase relationships between the subarrays still may be important for amplitude-comparison monopulse if the subarray voltages prior to the ADC are combined in the direction orthogonal to that of the measurement. For that combining to achieve the full signal power, the voltages must be coherent (i.e., have the same phase or nearly so).

1.3 Assumed Concepts

In writing any scientific work, one must make some assumptions about the concepts understood by the reader. While for electronic systems, one could start with Maxwell's equations and work from there, that is too ambitious and would add material that many readers already know. So, here are some basic concepts that the reader is assumed to have some understanding:

- That plane waves can be represented as $\exp[\pm j(\mathbf{k} \cdot \mathbf{x} \pm \omega t)]$, where $\mathbf{k}$ is the wave vector (essentially the direction of propagation), $\mathbf{x}$ is position, ω is radian frequency, and t is time. The direction of propagation is determined by the sign between the spatial and temporal terms. Plane waves are used to represent signals out of, propagating between, and into antennas when it is assumed that the observation point is in the far-field of the array. With this assumption, a Transmitted (Tx) planewave appears to be emanating from the phase center of the array, and the signal phase at a point on a Received (Rx) array is just $\mathbf{k} \cdot \mathbf{x}$, suppressing the Radio Frequency (RF) part ωt of the signal.
- That at baseband, narrowband signal and noise voltages can be represented as complex quantities, the real part being the in-phase voltage and the imaginary part being the quadrature-phase voltage. Because we are assuming narrowband signals (bandwidth is much less than the carrier frequency), we can ignore effects such as beam squint wherein the pointing angle of an array beam varies significantly across the signal bandwidth.
- That Additive White Gaussian Noise (AWGN) at baseband is complex, with zero-mean, equal variance, and uncorrelated real and imaginary parts. AWGN from different parts of an array or sampled at different times is uncorrelated sample to sample.

Chapter 2
Signal-to-Noise Ratios, the Relationship Between Signals-in-Space and Received Voltages, and AESA Gain

Abstract The performance of monopulse systems depends on the signal-to-noise ratio of the samples that are used in the measurements. A brief discussion of the communications link margin equation and the monostatic radar range equation is presented. Then the relationship between signals in space characterized by electric fields and received voltages is discussed. This leads to the understanding of signal and noise combining in hardware circuitry, where the concept of conservation of energy is invoked to describe how passive power combiners must operate. Finally, the relative magnitudes of signal and noise in an AESA before digital sampling are used to justify the usual expression for the gain of an AESA.

Keywords Signal-to-noise ratio (SNR) · Monostatic radar · Antenna effective area · Noise power spectral density · Coherent integration · Radar cross section (RCS) · Additive White Gaussian Noise (AWGN) · Passive power combiner · AESA gain · Conservation of energy

2.1 Signal-to-Noise Ratios

The derivation of the SNR in a one-way communications link is done first. Then this result is extended to a monostatic (i.e., the transmitter Tx and receiver Rx are collocated and use the same antenna for both) two-way radar.

Consider a transmitter with isotropic radiated power of P_{Tx} [W].[1] The Tx antenna focuses this energy in the direction of the receiver by applying a dimensionless gain of G_{Tx}. For this idealized calculation, we usually use the peak gain of the AESA, but in reality, rarely is the antenna pointed directly at the target, and one must be careful to reduce the peak power by the actual gain in the true direction of the line-of-sight. So, the peak power in the line-of-sight direction is $P_{Tx}G_{Tx}$. At the receiver's array face, the power per unit area is $P_{Tx}G_{Tx}/4\pi R^2$, where R is the distance from Tx to Rx. The Rx antenna has an effective area A_e that intercepts this power per unit area,

---

[1] Square brackets that follow a symbol indicate its units. We use SI units of Watts [W], Hertz [Hz], seconds [s], and meters [m] in these equations.

so the Rx signal power is $P_{Tx}G_{Tx}A_e/4\pi R^2$. The effective area of an array is related to its gain by the well-known formula [derived in a straightforward manner in Dana (2019)],

$$G_{Rx} = \frac{4\pi A_e}{\lambda^2},\tag{2.1}$$

where λ is the wavelength of the RF signal. Inserting this into the Rx power equation, the Rx signal power is

$$P_{Rx} = \frac{P_{Tx}G_{Tx}A_e}{4\pi R^2} = \frac{P_{Tx}G_{Tx}G_{Rx}}{(4\pi)^2}\left(\frac{\lambda}{R}\right)^2.$$

The noise power spectral density in the receiver is just $N_F k_B T_0$, where N_F is the noise factor (or noise figure if reported in decibels), Boltzmann's constant $k_B = 1.380649 \times 10^{-23}$ [W/Hz - $^\circ$K], and $T_0 = 290\,^\circ$K, room temperature. Then the noise power in the bandwidth of one voltage sample B_1 [Hz] is

$$P_N = N_F k_B T_0 B_1,\tag{2.2}$$

and the one-sample SNR at the receiver of a one-way link is

$$\frac{S}{N} = \frac{P_{Rx}}{P_N} = \frac{P_{Tx}G_{Tx}G_{Rx}}{(4\pi)^2 N_F k_B T_0 B_1}\left(\frac{\lambda}{R}\right)^2.$$

The systems engineer trying to make monopulse work to a required accuracy usually has very little control over the terms in this equation. As we shall see, the monopulse error decreases as the square root of SNR. The frequency and its wavelength usually are set by Government regulation for the type of system, and the maximum range is set by the operational concept. The noise power is set by physics unless one goes to extraordinary (read that expensive) ends; even then the noise temperature cannot be changed by much relative to absolute zero without cryogenics. Similarly, the Tx power amplifier contribution P_{Tx} is often limited by a system power budget, and the AESA gains are fixed by array size (in wavelengths). But the engineer has control over the signal bandwidth, and coherent integration of N_{CI} voltage samples makes the bandwidth in this expression smaller by the number of samples integrated ($B_{CI} = B_1/N_{CI}$). Hence, coherent integration directly increases the SNR. Of course, we are stretching the term "monopulse" a bit with coherent integration, and it takes longer to make a measurement. But this may be the best way to improve performance at the lowest cost to the system. The multi-sample SNR becomes

$$\frac{S}{N} = \frac{P_{Rx}}{P_N}$$

$$= \frac{P_{Tx}G_{Tx}G_{Rx}N_{CI}}{(4\pi)^2 N_F k_B T_0 B_1} \left(\frac{\lambda}{R}\right)^2 \text{(One-Way, Multi-Sample Signal-to-Noise Ratio)}.$$

Still, the signal bandwidth must be decreased by a factor of 4 (6 dB increase in SNR) to decrease the monopulse error by a factor of 2.

For a monopulse radar tracking a target, the transmitted power incident on the target is $P_{Tx}G_{Tx}/4\pi R^2$. The target intercepts this power and scatters it back toward the transmitter with an effective area of the radar cross section (RCS) σ [m^2], and that "transmitted power" scattered back from the target is $P_{Tx}G_{Tx}\sigma/4\pi R^2$. Then at the Rx the power per unit area incident on the antenna is again diminished by the $1/4\pi R^2$ factor and is collected by the antenna of effective area[2] $A_e = \lambda^2 G_{Rx}/4\pi$. Thus, the received power is

$$P_{Rx} = \frac{P_{Tx}G_{Tx}\sigma}{4\pi R^2} \frac{1}{4\pi R^2} \frac{\lambda^2 G_{Rx}}{4\pi} = \frac{P_{Tx}G_{Tx}G_{Rx}}{(4\pi)^3} \frac{\sigma\lambda^2}{R^4}.$$

The receiver noise power is characterized by a similar expression as was used for one-way propagation, but in this case the transmitted pulsewidth τ determines the bandwidth of a voltage sample ($B_1 \simeq 1/\tau$). Assuming that we must coherently integrate N_{CI} pulses to achieve the needed monopulse accuracy, the two-way SNR after integration is[3]

$$\frac{S}{N} = \frac{P_{Tx}G_{Tx}G_{Rx}N_{CI}\tau}{(4\pi)^3 N_F k_B T_0} \frac{\sigma\lambda^2}{R^4} \text{(Two-Way, Multi-Pulse Signal-to-Noise Ratio).} \quad (2.3)$$

Now the systems engineer has two degrees of freedom to increase the SNR without major systems changes: increasing the pulsewidth and its corresponding matched filter (another form of coherent integration) and the number of pulses coherently integrated.

---

[2] By the principle of reciprocity in electrodynamics, the gain of the aperture is the same on transmit and receive. But this assumes that the paths in the AESA are the same both ways, which is not true as the Tx side has a power amplifier whereas the Rx side may have a low-noise amplifier, and these two are distinctly different. For monopulse in particular, the Tx and Rx arrays are different as the full array is used on Tx but subarrays are used on Rx.

[3] It may not be immediately obvious that the SNR in this expression is dimensionless, as it necessarily must be. The RCS has units of area, so the ratio $\sigma\lambda/R$ is dimensionless. The noise temperature term $k_B T_0$ has units of energy or power×sec or power per Hertz. Thus $k_B T_0 B_1$ has units of power, cancelling the units of the transmitted power.

2.2   Signals in Space and in Electronics and Noise

In formulating these expressions, we have ignored the fact that in free space the power of a propagating RF signal is contained in its electric and magnetic fields, and similarly in the analog circuitry of the Tx or Rx the signal is represented by voltages or currents. So, for those not familiar with these electromagnetic slights of hand, further explanation may be needed. In free space, characterizing the electric field $\mathbf{E}(t)$ is sufficient to describe the propagating signal per Maxwell's equations. That the units of $\mathbf{E}(t)$ are expressed as volts per meter conjures up an image of a tiny electrical engineer with a voltmeter and ruler. But a better way to understand these units is to recognize that $\|\mathbf{E}(t)\|^2$ has units volts^2/m^2. The power in a circuit with voltage V and impedance Z is $P = V^2/Z$. So for free space, the impedance $Z_0 \simeq 120\pi$ [ohms], and the power of the propagating signal is $\|\mathbf{E}(t)\|^2/Z_0$ that represents the power per unit area incident on the aperture of an AESA or on a radar target. Within the Tx or Rx circuitry, when we take the power to be the voltage amplitude squared, we are tacitly assuming a 1 Ω impedance.

Now a tricky part. In the monopulse developments to follow, we consider breaking the AESA into two or four subarrays, and separately receiving the signal from each. But the SNR expressions above incorporate the gain of the full array and compute the signal power from the full array. So, what is the signal power out of each subarray? Of course, it must be equally divided between the subarrays, so the signal power out of a subarray should be, without too much thought, either $P_S/2$ or $P_S/4$ for two or four subarrays, respectively.

Let us do a little Gedanken experiment on this insight. The amplitude of the signal voltage is the square root of the power, so for 2 subarrays each has a signal amplitude of $\sqrt{P_S/2}$. If these combined coherently by adding voltages (i.e., if they have nearly the same phase, as they should if the AoA is in the main beam), then the combined signal amplitude should be $2\sqrt{P_S/2}$, which has a power of $2P_S$. Oops, energy is not conserved, a very embarrassing situation for a physicist! Before submitting a patent on this, we need to consider how analog voltages are combined in actual hardware, and the best way to see how this must happen is to consider noise voltages. The noise power in the analog circuitry is given by the expression in Eq. (2.2). This power is proportional to the temperature, and that must be the same on both sides of the combiner.[4]

So, let the Additive White Gaussian Noise (AWGN) baseband voltages in the front-end bandwidth before pairwise combining be represented as

---

[4]Voltage combining is done in an electrical circuit made of metal, a good conductor of electricity and of heat, so there cannot be a steady state temperature difference between the input side and the output side.

$$n_1 = x_1 + jy_1$$
$$n_2 = x_2 + jy_2,$$

where the statistics of AWGN[5] require that

$$\langle x_1 \rangle = \langle x_2 \rangle = \langle y_1 \rangle = \langle y_2 \rangle = 0$$
$$\langle x_1 x_2 \rangle = \langle x_1 \rangle \langle x_2 \rangle = \langle y_1 y_2 \rangle = \langle y_1 \rangle \langle y_2 \rangle = 0$$
$$\langle x_1^2 \rangle = \langle y_1^2 \rangle = \langle x_2^2 \rangle = \langle y_2^2 \rangle = \sigma_N^2$$
$$\langle n_1 n_1 \rangle = \langle n_2 n_2 \rangle = \langle n_1 n_2 \rangle = 0$$
$$\langle n_1 n_1^* \rangle = \langle n_2 n_2^* \rangle = 2\sigma_N^2.$$

Consider that these voltages are scaled in the combining hardware by a factor α. Then the combined noise voltage is

$$n_{12} = \alpha(n_1 + n_2),$$

and the combined mean noise power is

$$P_N = \langle |n_{12}|^2 \rangle = \alpha^2 \langle |n_1|^2 \rangle + \alpha^2 \langle |n_2|^2 \rangle = \alpha^2 4\sigma_N^2,$$

where the brackets $\langle x \rangle$ denote ensemble averages that for ergodic process are achieved by sufficiently long time averages. To ensure that this is the same as the noise powers of the inputs before combining, it is necessary that $\alpha = 1/\sqrt{2}$. This conserves noise energy and keeps the temperature the same on both sides of the combiner (per Eq. (2.2)).[6]

This also must be the same scaling of the signal voltages in the combiner. Thus, the combined signal amplitude and power of the subarray voltages is

$$a_{12} = 2\alpha\sqrt{P_S/2}$$
$$P_{12} = a_{12}^2 = 2\alpha^2 P_S = P_S,$$

which conserves signal energy also.

This is referred to as a passive power combiner (passive because no amplification is applied), and it works with pairs of input voltages. So, for our example 1024

---

[5]Zero mean, complex Gaussian with uncorrelated real and imaginary parts and uncorrelated sample to sample.

[6]The thermal balance of an element is not maintained just by RF signals and noise. Indeed, metallic contact with adjacent elements and with a substrate may well play a larger part. We maintain this convention for mathematical convenience, at least for ideal cases. But instantaneously, noise can vary over a wide range so one must be careful to distinguish between what happens in a sample versus what happens on the average over a long time period.

element array, there must be 10 layers of passive power combiners in the full array, each applying the same factor $\alpha = 1/\sqrt{2}$ to the input voltages to get the single output, thereby preserving the noise temperature before and after combining and preserving signal energy before and after also.

It is important to note that this analysis is for analog voltage combining. It does not apply when digital representations of the voltages are combined.

2.3   Gain of an AESA

Weighting of elements, whether in Tx or Rx mode, is often done to control the sidelobe levels of the antenna gain versus angle-of-arrival. So, considering the Rx mode, the signal s_m plus noise n_m voltages out of the mth element can be written

$$v_m = w_m(s_m + n_m),$$

where w_m is the weight applied to that element. The outputs of all the elements in the array are summed together with passive power combiners. After writing down the pair-by-pair summed voltages for $\log_2 N_e$ levels, assuming that the number of elements N_e is a power of 2, which is what we will do, one can see that the signal plus noise voltage output of the array is

$$v_S + v_N = \alpha^{\log_2(N_e)} \sum_{m=1}^{N_e} w_m(s_m + n_m).$$

Using the value of α from above, the leading term is just $\alpha^{\log_2(N_e)} = 1/\sqrt{N_e}$.

The gain of an electronic device is defined as the SNR at the output relative to that at the input. The noise power at the output is

$$P_N = \left\langle |v_N|^2 \right\rangle = \frac{P_{N,1}}{N_e} \sum_{m=1}^{N_e} |w_m|^2,$$

where $P_{N,\,1}$ is the AWGN power of an element before weighting. Assuming that the signal voltages out of each element have the same phase, which will give the maximum array gain, the signal power output is

$$P_S = \frac{P_{S,1}}{N_e} \left| \sum_{m=1}^{N_e} w_m \sqrt{G_{e,m}} \right|^2,$$

where $\sqrt{G_{e,m}}$ is the gain of the mth element that we will assume is the same for every element. This factor is not included in $P_{S,\,1}$, meaning that all Rx antenna contributions have been removed from it. We will need to remember this later when simulating monopulse performance.

The gain of the AESA in the Rx mode is then $(P_S/P_N)/(P_{S,\,1}/P_{N,\,1})$, which gives the usual result for AESAs:

$$
\boxed{G_{AESA} = G_e \frac{\left| \sum_{m=1}^{N_e} w_m \right|^2}{\sum_{m=1}^{N_e} |w_m|^2}.}
$$

By the principle of reciprocity, this is also the gain of the AESA in the Tx mode. As this is a "first principles" expression, we put it in a box. For an ideal antenna, the gain of an element is computed from the formula in Eq. (2.1). For an element with spacing from nearest neighbors of $\lambda/2$ and ideal effective area of $\lambda^2/4$, gain is just $G_e = \pi$. The maximum gain of our 32×32 element array with uniform weighting is 1024π or 35.07 dB.

Note that the AESA gain expression is independent of our assumption that the element output voltages are summed pairwise in passive power combiners. For example, it holds even if there happens to be a radio behind each element so that voltages are summed numerically.

In developments below, we set the noise gain of the array to unity. That is, we normalize the weights so that

$$
\sum_{m=1}^{N_e} |w_m|^2 = 1. \tag{2.4}
$$

This normalization means that there is one less thing to compute in a simulation, and fewer required calculations in a simulation is often an advantage in the time one must wait for an answer.

This normalization becomes problematic, however, when we consider nonideal behavior from quantization or failed elements or Vector Modulator (VM) noise described in Chap. 6. (It is a VM that applies the amplitude and phase to an element.) What happens when these effects modify the weights? Does the array magically renormalize? Obviously not, so we depart from this convention in nonideal cases, and account for non-unity noise gain in the calculation of the SNR gain of the nonideal array. The AWGN of the received voltages also experiences both its usual fluctuations and those caused by instantaneous fluctuations of weights that in turn are caused by VM noise, for example.

So, what about conservation of energy? We rely on the assumed thermal conductivity of the in situ elements to maintain a constant RF noise temperature. However, thermal conductivity and equilibrium does not happen instantaneously, and fluctuations in the weights and in the sum in Eq. (2.4) are physically realistic.

Chapter 3
Theory of Phase- and Amplitude-Comparison Monopulse

Abstract The ideas behind phase- and amplitude-comparison monopulse are simple physical descriptions that lead to equations that relate signal voltages to angular displacement measurements of the true angle of arrival from the AESA pointing direction. In the case of amplitude-comparison monopulse, these descriptions allow an exact calculation of the measurement error that is used to determine the squint angle and validate the simulations of these processes.

Keywords Phase-comparison monopulse · Amplitude-comparison monopulse · Subarrays · Beamwidth · Squint angle · Rician distribution

3.1 Relationship Between AESA Pointing Angles and Direction of Propagation

The direction of propagation of a planewave is characterized by the wave vector $\mathbf{k}$, and the phase of the planewave at any point $\mathbf{x}$ is represented by the dot product $\mathbf{k} \cdot \mathbf{x}$, ignoring the time contribution ωt. So, in a rectangular coordinate system centered on an AESA and aligned with the sides of the array, the wave vector of a transmitted planewave is illustrated in Fig. 3.1, where the boresight of the array, perpendicular to its face, is along the z axis, and the array is in the $z = 0$ plane. This vector also can be represented in the spherical-polar coordinates of the array in terms of its magnitude $\|\mathbf{k}\| = k_0 = 2\pi/\lambda$, elevation angle θ, and azimuthal angle φ.

The $x - y - z$ components of the wave vector are then

$$\mathbf{k} = k_0(\cos\varphi\sin\theta\,\widehat{\mathbf{x}} + \sin\varphi\sin\theta\,\widehat{\mathbf{y}} + \cos\theta\,\widehat{\mathbf{z}}),$$

where $\widehat{\mathbf{x}}$, $\widehat{\mathbf{y}}$, and $\widehat{\mathbf{z}}$ are unit vectors in the three orthogonal directions. A planewave received by the array from that same direction has the wave vector $-\mathbf{k}$. In further developments, we sometimes refer to a unit vector $\mathbf{u} = \mathbf{k}/k_0$.

For an outgoing wave, the array is pointed in the elevation and azimuthal directions θ_0, φ_0, respectively:

Fig. 3.1 Wave vector in an
AESA rectangular
coordinate system

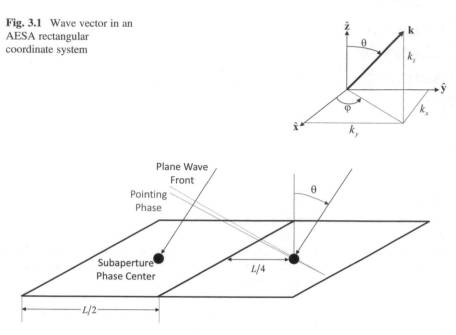

Fig. 3.2 Phase-comparison monopulse with two subarrays

$$\mathbf{k}_0 = k_0 (\cos \varphi_0 \sin \theta_0 \, \widehat{\mathbf{x}} + \, \sin \varphi_0 \sin \theta_0 \, \widehat{\mathbf{y}} + \, \cos \theta_0 \, \widehat{\mathbf{z}}).$$

To receive a signal from that direction, the array is pointed in the direction $-\mathbf{k}_0$.

Later we will assume that transmitter or radar target are in the far field of the array, so we can represent the signals in space as planewaves.

3.2   Phase-Comparison Monopulse Measurements in One Dimension

Consider an AESA split into two subarrays, as illustrated in Fig. 3.2. The length of the total array is L, and each subarray has length $L/2$ assuming that the element spacing across the subarray boundary remains the same as that in the interior of the subarray. If the full array is to have a significant gain, then L will encompass many wavelengths, and if the array is not to suffer from grating lobes, the element spacing will be no larger than $\lambda/2$ in both dimensions. We further assume that the summing network within the subarray is such that the phase center[1] of the subarray, indicated

---

[1] On transmit, the phase center is the point from which a spherical wave appears to be emanating when observed in the far field. On receive, an analogous definition of the phase center is the point where the phase of the array combined voltage is equal to that of the incident wave at that point, ignoring the imparted phase of the summing network. For our purposes, the phase center is the geometric center of the array or subarray.

by the dots, is at its geometric center. This requires that the phase distance from each element to the summing point is the same.

If the array does indeed have a narrow beamwidth and significant gain, then it must be many wavelengths in size, and $L > > \lambda$. So, the extra phase at the left subarray of the arriving planewave also must be many wavelengths and is ambiguous unless the AoA is very small. But if the array is pointed at or near the direction of $-\mathbf{k}$, then the phase difference that is important is difference between the pointing direction phase ramp across the array (dashed red line) and that of the incident signal (black dashed line). One can see that if the AoA is within the first nulls of the array gain measured from the peak, then the phase difference between the two subarrays is indeed unambiguous, which is how the main beam is formed.

Consider first the one-dimensional (1-D) case illustrated in the figure. The center of the $x - y - z$ coordinate system is at the geometric center of the full array that sits in the $z = 0$ plane. Then the vector locations of the left and right subarray phase centers in terms of the $x-$direction full array length L_x is

$$\mathbf{x}_L = -(L_x/4)\widehat{\mathbf{x}} \quad \mathbf{x}_R = +(L_x/4)\widehat{\mathbf{x}}.$$

Now consider a planewave incident on the array at an angle from boresight (the direction normal to the face of the array along the z axis) of θ. In this 1-D case, the azimuthal AoA and pointing angles are both zero. Then there is a simple relationship between the received signal phase of the right subarray relative to that of the left subarray:

$$\Delta\phi = -(\mathbf{k} - \mathbf{k}_0).(\mathbf{x}_R - \mathbf{x}_L) = -(k_x - k_{0x})L/2$$
$$= -k_0(\sin\theta - \sin\theta_0)L/2,$$

where θ_0 is the pointing direction elevation angle, and the leading minus sign is there because this is an arriving planewave.

But in an antenna, we do not get the received signal phase directly. What we get is the signal plus noise voltage of the left and right subarrays. For the moment, we will ignore the noise, and write the received voltages from the left and right subarrays as

$$V_R = \text{Voltage from Right Subarray} = a_0/2$$
$$V_L = \text{Voltage from Left Subarray} = (a_0/2)\exp(-j\Delta\phi),$$

ignoring the time contribution assuming that both voltages are captured at the same time. Here a_0 is the signal amplitude if it were received by the full array, so $a_0/2$ is the signal amplitude of a subarray.

We now form sum and differences voltages digitally per the block diagram in Fig. 1.1, so we do not need the $1/\sqrt{2}$ scaling (as the monopulse measurements are formed from the ratio Δ/Σ, any common scaling will drop out anyway):

$$\Sigma = V_R + V_L = a_0 \exp\left(-j\Delta\phi/2\right) \frac{\left[\exp\left(+j\Delta\phi/2\right) + \exp\left(-j\Delta\phi/2\right)\right]}{2}$$

$$= a_0 \exp\left(-j\Delta\phi/2\right) \cos\left(\Delta\phi/2\right)$$

$$\Delta = V_R - V_L = a_0 \exp\left(-j\Delta\phi/2\right) \frac{\left[\exp\left(+j\Delta\phi/2\right) - \exp\left(-j\Delta\phi/2\right)\right]}{2}$$

$$= j\Sigma \tan\left(\Delta\phi/2\right).$$

Taking the ratio of these two and inverting the tangent function, we get the phase difference:

$$\widetilde{m} = \tan^{-1}\left[\text{Re}\left(\frac{j\Delta}{\Sigma}\right)\right] = -\frac{\Delta\phi}{2},$$

where $\widetilde{m}$ denotes the phase-comparison measurement. Inverting this equation for the measured elevation angle of arrival,

$$\widetilde{\theta} = \sin^{-1}\left(\frac{\widetilde{m}}{k_0 L/4} + \sin\theta_0\right).$$

3.3 Phase-Comparison Monopulse Measurements in Two Dimensions

Now consider phase-comparison monopulse in two dimensions. Our example full 32×32 element array is broken into four 16×16 element subarrays, as illustrated in Fig. 3.3. The red lines indicate the boundaries of the subarrays, the red dots indicate the phase centers of the subarrays, and the numbers correspond to the four output voltages. The sum and difference voltages are formed digitally as

$$\Sigma = V_1 + V_2 + V_3 + V_4$$
$$\Delta_x = (V_2 + V_4) - (V_1 + V_3)$$
$$\Delta_y = (V_1 + V_2) - (V_3 + V_4).$$

Once we combine the voltages, the phase centers shift to the lines dividing the subarrays. So, for the two directions (Horizontal Left HL and Right HR and Vertical Lower VL and Upper VU, relative to the figure), the phase centers are at the positions

$$\mathbf{x}_{HL} = -(L_x/4)\widehat{\mathbf{x}} \quad \mathbf{x}_{HR} = +(L_x/4)\widehat{\mathbf{x}}$$
$$\mathbf{x}_{VL} = -(L_y/4)\widehat{\mathbf{y}} \quad \mathbf{x}_{VU} = +(L_y/4)\widehat{\mathbf{y}},$$

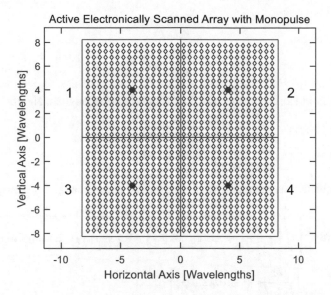

Fig. 3.3 A 32×32 element AESA broken into four subarrays for phase-comparison monopulse

where L_y is the length of the full array in the y–dimension. Then the phase differences in the two dimensions are

$$\Delta\phi_x = -(\mathbf{k} - \mathbf{k}_0).(\mathbf{x}_{HR} - \mathbf{x}_{HL}) = -(k_x - k_{0x})L_x/2$$
$$= -k_0(\sin\theta\cos\varphi \quad \sin\theta_0\cos\varphi_0)l_x/2$$
$$\Delta\phi_y = -(\mathbf{k} - \mathbf{k}_0).(\mathbf{x}_{VU} - \mathbf{x}_{VL}) = -(k_y - k_{0y})L_y/2$$
$$= -k_0(\sin\theta\sin\varphi - \sin\theta_0\sin\varphi_0)L_y/2.$$

From these expressions, the two "first principles" monopulse measurements are:

$$\boxed{\begin{aligned} \widetilde{m}_x &= \tan^{-1}\left[\operatorname{Re}\left(\frac{j\Delta_x}{\Sigma}\right)\right] = -\frac{\Delta\phi_x}{2} \\ \widetilde{m}_y &= \tan^{-1}\left[\operatorname{Re}\left(\frac{j\Delta_y}{\Sigma}\right)\right] = -\frac{\Delta\phi_y}{2} \end{aligned}} \tag{3.1}$$

Note that these expressions are the same as that for the 1-D case, but are applied in the two directions. Using the expressions for the delta-phase as functions of the unknown AoA k-vector components, denoted as $\widetilde{k}_x$ and $\widetilde{k}_y$ because they are estimates, the measured AoA components in terms of the pointing vector components are

$$\widetilde{k}_x = \frac{\widetilde{m}_x}{k_0 L_x/4} + k_{0,x}$$

$$\widetilde{k}_y = \frac{\widetilde{m}_y}{k_0 L_y/4} + k_{0,y}. \tag{3.2}$$

The errors in these estimates are the difference between them and the truth:

$$\varepsilon_x = \widetilde{k}_x - k_x$$

$$\varepsilon_y = \widetilde{k}_y - k_y. \tag{3.3}$$

For completeness, the measured elevation and azimuth angles are, respectively,

$$\widetilde{\theta} = \sin^{-1}\left[\sqrt{\left(\frac{\widetilde{m}_x}{k_0 L_x/4} + \sin\theta_0 \cos\varphi_0\right)^2 + \left(\frac{\widetilde{m}_y}{k_0 L_y/4} + \sin\theta_0 \sin\varphi_0\right)^2}\right]$$

$$\widetilde{\varphi} = \tan^{-1}\left[\frac{\dfrac{\widetilde{m}_y}{k_0 L_y/4} + \sin\theta_0 \sin\varphi_0}{\dfrac{\widetilde{m}_x}{k_0 L_x/4} + \sin\theta_0 \cos\varphi_0}\right].$$

In further developments, when we discuss phase-comparison monopulse errors, we are referring to the errors in $\widetilde{k}_x$ and $\widetilde{k}_y$, not the errors in the reported elevation and azimuthal angles.

Equation (3.1) relates phase-comparison measurements made from voltages to the AoA of the incident signal and is a "first principles" expression, hence the box, even though it is derived from voltage expressions, albeit in a straightforward manner. Later we will see how these equations relate to the signal and noise voltages at the output of elements in a simulation of this measurement process using an AESA with amplitude taper and failed elements, for example.

3.4 Amplitude-Comparison Monopulse Measurements in One Dimension

Amplitude-comparison monopulse is inherently a one-dimensional process as different pointing angles are applied to each subarray, and one cannot simultaneously apply two different phases to an element. So, to get AoA in two dimensions, sequential measurements must be made with a pointing angle change made to the elements in between. This puts amplitude-comparison monopulse at an immediate disadvantage relative to phase-comparison monopulse that can be done in two

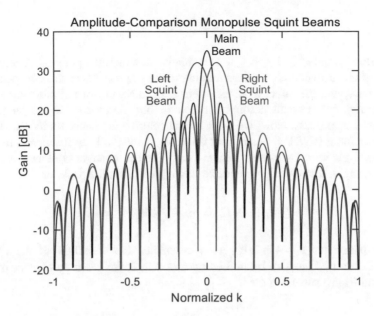

Fig. 3.4 Squinted beams of amplitude-comparison monopulse

angular directions simultaneously. We include it in our discussion for completeness, to show why phase-comparison monopulse is the preferred method in AESAs, and because it allows an analytic error analysis that can be used to validate our simulation of both methods.

The concept of amplitude-comparison monopulse is illustrated in Fig. 3.4, where three beam profiles are plotted versus normalized k (either the $\widehat{\mathbf{k}}_x$ or $\widehat{\mathbf{k}}_y$ components of the unit vector $\widehat{\mathbf{k}}$). The black profile is the main beam gain of the full array, and the blue and magenta profiles are two squinted beams each pointed away from the main beam by half the squint angle θ_S, the angular separation of the squint beam pointing directions. As these are formed from two subarrays that split the full array in half, they have 3-dB lower peak gain and beamwidths twice as large as that of the main beam. The true AoA should be somewhere between the two squint beam peaks. By comparing the signal amplitudes from the two subarrays, one can measure the angular distance the true AoA is from the pointing direction of the main beam. Because an element can only apply one pointing phase at a time to the incident signal, the squint beams and the main beam are not simultaneous.

For mathematical convenience in the following developments, we assume that the beam profiles have a Gaussian shape in k-space. That is, the one-dimensional power beam profile of the main beam is

$$G(k) = \exp\left[-\beta(k - k_P)^2\right],$$

where normalized k^2 $(-1 \le k \le +1)$ is relative to a k_P is the pointing direction. Of course, there are a couple of things missing in this expression, such as the peak gain being unity, and the effect of foreshortening on the gain for pointing angle away from unity, but we will account for that in our simulation of the monopulse measurement process. Now, we want to characterize the beam by its Full-Width, Half-Maximum (FWHM) beamwidth is denoted as Θ_0 to keep it separate from the pointing angle we subsequently denote as θ_0. Then at an angle $\Theta_0/2$ away from the pointing angle, the power gain must be down 3 dB from the peak, or

$$G(k_P \pm \Theta_0/2) = \exp\left[-\beta\Theta_0^2/4\right] = \frac{1}{2}.$$

The parameter $\beta = 4\ln 2/\Theta_0^2$ then completes the definition of the FWHM beamwidth for a Gaussian beam. Letting $\alpha = 4\ln 2$, the beam profiles for the two (Left and Right) subarrays are

$$G_L(k_A) = \exp\left[-\alpha\left(\frac{k_A - k_P - k_S/2}{\Theta_0}\right)^2\right]$$

$$G_R(k_A) = \exp\left[-\alpha\left(\frac{k_A - k_P + k_S/2}{\Theta_0}\right)^2\right],$$

where k_A is the true AoA (or direction of arrival) and k_S is squint beam separation.

To get the AoA from these expressions, we take the natural logarithm of the subarray voltage amplitudes v_R and v_L (the square root of the power gains given above) that have twice the beamwidth of the full array, ignoring noise and the signal power:

$$\Delta = \ln\left[\frac{v_R}{v_L}\right] = \ln\left[\sqrt{\frac{G_R}{G_L}}\right] = \ln\left[\frac{\exp\left[-\left(\frac{\alpha}{2}\right)\left(\frac{k_A-k_P+k_S/2}{2\Theta_0}\right)^2\right]}{\exp\left[-\left(\frac{\alpha}{2}\right)\left(\frac{k_A-k_P-k_S/2}{2\Theta_0}\right)^2\right]}\right] = \frac{\alpha\theta_S(k_A - k_P)}{\Theta_0^2}.$$

Solving for k_A, we get the amplitude-comparison monopulse "first principles" equation for the measured angle of arrival in terms of the pointing direction:

---

[2] In one dimension the normalized k is equal to $\sin\theta$, where the angle θ is measured from boresight. We assume here that the beamwidth of the array is small enough that $\sin\theta \simeq \theta$, and $k = \theta$. For mathematical convenience we then extend the domain of k to $-\infty < k < +\infty$ assuming the beamwidth is small compared to one radian.

$$\boxed{\widetilde{k}_A = k_P + \frac{\Theta_0^2}{k_S \ln(2)} \ln\left[\frac{v_R}{v_L}\right].}$$ (3.4)

The squint angle enters this expression in two ways. First, there is the explicit term Θ_0/k_S in the measurement error $\varepsilon = \widetilde{k}_A - k_A$. This would seem to indicate that the squint angle should be made as large as practical. But the signal arrives after being attenuated by the squint beam profiles that get smaller exponentially as the true AoA gets farther from the pointing direction. So, there must be an optimum squint angle. This will be computed next.

3.5 Amplitude-Comparison Monopulse Theory in One Dimension

The first principles expression in Eq. (3.4) can be separated into two terms involving independent random variables. That is, rewrite the logarithmic term as

$$\ln\left[\frac{v_R}{v_L}\right] = \ln\left[\frac{v_R}{\sigma_N}\right] - \ln\left[\frac{v_L}{\sigma_N}\right],$$

where σ_N is the noise voltage standard deviation in either the in-phase or quadrature phase channels.[3] We know two things about these amplitudes that apply here. First, the AWGN samples in the two subarray outputs are uncorrelated, zero-mean, normally distributed random variables. Then, the signal plus noise voltages have a Rician distribution (Papoulis and Pillai 2002). That is, the probability density function of either of the two voltage amplitudes is

$$f(v,a) = \frac{v}{\sigma_N^2} \exp\left[-\frac{v^2 + a^2}{2\sigma_N^2}\right] I_0\left[\frac{va}{\sigma_N^2}\right],$$

where I_0 is the modified Bessel function of first kind and zeroth order, v is the amplitude of the signal plus noise voltage, and a is the amplitude of signal contribution to that voltage.

The mean and standard deviation of the measurement error $\varepsilon = \widehat{\theta}_A - \theta_A$ then involves the two moments of the natural logarithm:

---

[3]This normalization is done to keep the argument of the logarithms dimensionless. Because $\ln(\sigma_N)$ adds and subtracts from the other logarithmic expressions, it has no effect on the result.

$$\langle \ln^n(v/\sigma_N)\rangle = \int_0^\infty \ln^n(v/\sigma_N)f(v,a)dv$$

$$= \frac{1}{\sigma_N}\int_0^\infty \ln^n(v/\sigma_N)\frac{v}{\sigma_N}\exp[-\frac{v^2+a^2}{2\sigma_N^2}]I_0[\frac{vthickmathspacea}{\sigma_N^2}]dv.$$

It turns out that there are formulas for the integral that involve infinite sums of poly gamma functions, but these forms can be cantankerous to evaluate. However, the integral is easily evaluated numerically once one notices that the scaled Bessel function expression $e^{-x}I_0(x)$ does not blow-up faster than the leading exponential term reduces the integrand for large arguments. So, upon changing variables to $\xi = v/\sigma_N$ and writing the integrand in terms of the scaled Bessel function, the integrals to be evaluated become

$$\langle \ln^n(v/\sigma_N)\rangle = 2\int_0^\infty \ln^n(\xi)\xi\exp\left[-\left(\sqrt{\gamma}-\xi\right)^2\right]e^{-2\xi\sqrt{\gamma}}I_0\left[2\xi\sqrt{\gamma}\right]d\xi,$$

where $\gamma = a^2/2\sigma_N^2$ is the signal-to-noise ratio of a subarray in this case.

With knowledge of what is to follow, we select a signal-to-noise ratio of 30 dB from the full array, so γ in this expression is 1000/2 modified by the subarray beam profiles, and we look for the optimum squint angle. First, however, we must define the received signal powers from the two squinted beams that include the effects of the reduction in gain as the squint angle is increased and the reduction of the SNR for a subarray versus the full array:

$$P_L = \frac{P_S}{2}\exp\left[-4\ln 2\left(\frac{k_A - k_P + k_S/2}{2\Theta_0}\right)^2\right]$$

$$P_R = \frac{P_S}{2}\exp\left[-4\ln 2\left(\frac{k_A - k_P - k_S/2}{2\Theta_0}\right)^2\right].$$

Assuming perfect pointing, $k_P = k_A$ in these expressions, and the amplitude-comparison monopulse error comes from Eq. (3.4) with the gain profiles included is

$$\varepsilon = \widetilde{k}_A - k_A = \frac{\Theta_0^2}{\theta_S\ln(2)}\left[\ln(v_R) - \ln(v_L)\right] - k_A.$$

The ensemble mean-square error is computed as

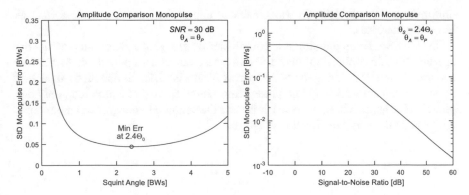

Fig. 3.5 Amplitude-comparison monopulse StD error versus squint angle and SNR

$$\langle \varepsilon^2 \rangle = \left(\frac{\Theta_0^2}{k_S \ln(2)} \right)^2 \left[\langle \ln^2(v_R) \rangle - 2 \langle \ln(v_R) \rangle \langle \ln(v_L) \rangle + \langle \ln^2(v_R) \rangle \right]$$

$$- 2 \frac{\Theta_0^2 k_A}{k_S \ln(2)} \left[\langle \ln(v_R) \rangle - \langle \ln(v_L) \rangle \right] + k_A^2,$$

where the ensemble expectations $\langle \varepsilon^n \rangle$ are averages over the Rician distribution. In further developments, we plot the Standard Deviation (StD)[4] of the error, $\sqrt{\langle \varepsilon^2 \rangle - \langle \varepsilon \rangle^2}$.

The amplitude-comparison monopulse error StD is plotted in the left frame of Fig. 3.5 versus squint angle. Here and in subsequent similar plots, angular measurements and errors are measured in beamwidths (BWs) of the full array pointed at boresight. The optimum squint angle that minimizes the error for a SNR of 30 dB is 2.4 beamwidths when the AoA is in the pointing direction of the array. Of course, as the problem has been formulated, this does not include the effects of beam broadening as the beam is pointed away from boresight, which goes as $1/\cos\theta$, but this is small (less than a factor of 1.155) for scan angles less than 30 degrees. The right frame of the figure is a plot of the error[5] versus SNR for the optimum squint angle and for the case where the main beam is pointed directly at the true AoA.

Although we could push the analytic model further, such as showing the measurement error versus pointing angle error ($k_P - k_A$), we must be aware that the idealized Gaussian beam profile differs from the $\sin^2(x)/x^2$ beam shape of a

---

[4]This error is most meaningful in an electronic system when the mean error is zero. If it is not, there may be a systematic bias in the measurements that must be corrected to minimize the error.

[5]These curves, however, are computed with another limitation of the problem—the numerical integration fails for SNR above 40 dB using the Matlab scaled Bessel function. By this value, however, it is clear that the error varies as the square root of the SNR, and the points above 40 dB are just the value at 40 dB scaled down by $\sqrt{10000/\gamma}$.

uniformly weighted AESA in ways that will impact the accuracy of the former as $(k_P - k_A)$ increases.

In an AESA, beam pointing is accomplished by applying a phase ramp across the aperture. In this application, two different phase ramps are applied across the two subarrays, thereby pointing the two subarray beams in different directions separated by the squint angle. As only one phase ramp can be applied at a time (i.e., only one phase can be applied in an element), it is clear that amplitude-comparison monopulse only can be done in one direction at a time.

Chapter 4
Sidelobe Control in Monopulse and Design Implications

Abstract In many applications, unwanted signals arriving away from the main beam of an AESA can mask the desired signal. Standard techniques for sidelobe control are discussed in the context of an AESA. Some limitations of an AESA have an impact on sidelobe control when simultaneously forming monopulse measurements and receiving the signal through the main beam.

Keywords Sidelobe control · Taylor weighting · Bayliss weighting · "Split-Taylor" weighting · Uniform weighting · Dynamic range

Sidelobe control is very important in many applications that can use an AESA. For example, in air traffic control or air defense radars one wants the target to be competing with AWGN, not ground clutter at the same range that enters through sidelobes. As the clutter can be much stronger than an aircraft return, the first line of defense against clutter is to keep the sidelobes as low as possible. In modern Pulsed-Doppler radars, Doppler processing also will separate the clutter from the target except for targets with slow radial velocity. Together, these two techniques may be sufficient. In an airborne weather radar where the strength of the return is presented to the pilots as varying colors on a plan position indicator (PPI), one does not want a ground return through ground-directed sidelobes to be mistaken as severe weather.

A standard technique for controlling the magnitude of array sidelobes is Taylor weighting (Taylor 1955). An example of Taylor weights and the resulting beam profile is shown in Fig. 4.1. The left frame has the weights of a 32 × 32 array along a middle row or column of elements. The right frame compares the Taylor-weighted beam (blue) with that for uniform weighting (black). We see that Taylor weighting results in a small loss in peak gain and broadening of the beamwidth. But the advantage of Taylor weighting is that the peak sidelobes are about 19 dB lower than those of the uniformly weighted $\sin^2(x)/x^2$ pattern.

Assuming Taylor weighting is applied both horizontally and vertically, each with column vectors of weights $\mathbf{w}_{e,H}$ and $\mathbf{w}_{e,V}$, the weight matrix of the 2-D array is just $W = \mathbf{w}_{e,V}\mathbf{w}_{e,H}^T$.

When we examine the performance of phase-comparison monopulse with quantization of the weight amplitude and phase, the question will arise as to the number of

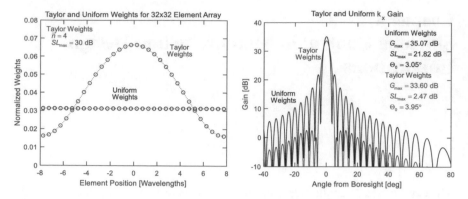

Fig. 4.1 Taylor and uniform weights for a 32×32 element array (left) and 1-D gains (right)

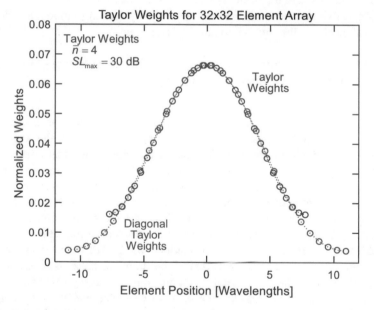

Fig. 4.2 Taylor weights straight across and diagonally across the 32×32 array

required bits for each. One can get an idea for amplitude by looking at the dynamic range of the weights, which varies depending on the direction we consider. Figure 4.2 shows the Taylor weights straight across a middle row or column of a 32×32 array (blue dots) and diagonally across the array (magenta dots). Along the diagonal, there are still 32 elements but these are separated by $\sqrt{2}\lambda/2$, of course. The weight dynamic range (*DR*) for the two cases and our 32×32 array is

$$DR_{Row} = 20 \log_{10} \left[\frac{\max(\mathbf{w}_{e,H})}{\min(\mathbf{w}_{e,H})} \right] = 12.27 \text{ dB}$$

$$DR_{Diag} = 20 \log_{10} \left[\frac{\max(W)}{\min(W)} \right] = 24.54 \text{ dB}$$

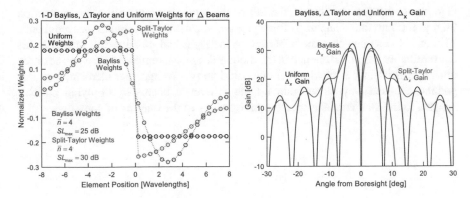

Fig. 4.3 Difference beam weights and resulting gain profiles

That these differ by a factor 2 in decibels, for the same weighting in each direction, makes sense because the weights of the corner elements are the square of the smallest 1-D weights. As the dynamic range of quantization is 6 dB per bit (voltages are quantized, and each bit corresponds to a factor of 2 in voltage or 6 dB in power), to capture the full range of Taylor weights for this square array requires 5 bits (although we could probably get away with just 4).

Bayliss (1968) developed difference beam weights based on the Taylor algorithm. These are plotted in the left frame (blue dots) of Fig. 4.3, along with two other examples of uniform weights (black dots) and "Split-Taylor" weights (magenta dots). The corresponding 1-D difference beam gains are plotted in the right frame using the same color scheme.

The Split-Taylor algorithm is a concession to the fact that for the simplest element design, only one weight and phase can be applied at a time. But simultaneous phase-comparison monopulse requires both sum and difference beams ideally with Taylor weights on the sum beam and Bayliss weights on the difference beams. One way around this is to use Taylor weights on both the sum and difference beams, but for the latter the sign of the weights is negative for one-half of the array (hence "Split-Taylor"). Sherman and Barton (2011, §7.4) introduce the idea of Split-Taylor weights, although they do not use this term.

A number of things are evident from these plots: (1) The Split-Taylor algorithm produces high sidelobes although not as high as uniform weighting does, and that its sidelobe structure does not exhibit deep nulls between the sidelobe peaks. This may be a disadvantage in the presence of jammers that can appear anywhere outside the main lobe. (2) For the Bayliss weight case (blue line), the peak sidelobes are below the range of the ordinate and do not appear on the right plot. (3) It should be clear from the weights plotted in the right frame that simultaneous 2-D difference beams are not possible with only one weight per element and Bayliss weights. However, when the aperture is split into four subarrays, the outputs of the subarrays can be alternately added and subtracted to form difference beams with the proper signs on the difference beam voltages when using any of the three difference beam weighting algorithms. (4) The weights on the left of the figure have the opposite convention used in the previous chapter to form phase-comparison monopulse measurements. That is, we assumed that the subarray on the right had positive weights. But this convention makes no difference in the monopulse measurement as long as one or the other is consistently applied. (5) If one could do two weights per element, then it is possible to used Bayliss weights for the difference beams by applying positive weights to the elements and then the minus signs to the outputs of subarrays.

Chapter 5
Ideal One-Way Monopulse Performance

Abstract One-way monopulse performance is described for the case where the receiving AESA does not transmit the signal, such as in a communications link AESA that tracks a moving transmitter or a bistatic passive radar receiver. In this chapter are descriptions of simulations of phase- and amplitude-comparison monopulse developed from first principles. That is, construction of signal plus noise voltages that form measurements in an ideal AESA where only AWGN corrupts the monopulse measurements. Some real-world effects are discussed in Chap. 6.

Keywords Phase-comparison monopulse simulation · Amplitude-comparison monopulse simulation · Monopulse performance · Mean error · Standard deviation (StD) error

There are a large number of cases that one could consider in quantifying performance, but only a limited number that are necessary for understanding that performance. To limit the possibilities, we consider a 32×32 element square array in the $z = 0$ plane with $\lambda/2$ element spacing. For most cases, the array is pointed at boresight, so the peak of the beam will be along the $z-$direction. Then we compute the monopulse performance in the $x-$ and $y-$directions at 9 true AoA positions represented by the dots in Fig. 5.1, but we only report the error at the four positions indicated as Peak, W, NW, and N, and in some cases only at the true AoA. The outer points are on the 3-dB beamwidth (FWHM) circle, and the beam is pointed at the peak position.

5.1 Simulation of Phase-Comparison Monopulse

Consider a plane-wave signal propagating in direction $-\mathbf{k}$ incident on a 32×32 element ESA (an AESA except without the active transmitter) separated into four subarrays each with an analog front end as illustrated in Fig. 1.1. The ESA is pointed

Fig. 5.1 Monopulse
measurement positions
relative to the pointed
beam peak

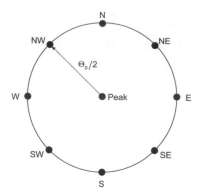

in the direction $+\mathbf{k}_0$ to cancel the AoA phase across the array, so the signal voltage
out of the array can be written as

$$v_S = a_0 g_e(\mathbf{k}) \sum_{m=1}^{N_e} w_m \exp\left[j(-\mathbf{k} + \mathbf{k}_0) \cdot \mathbf{x}_m \right], \qquad (5.1)$$

where $g_e(\mathbf{k})$ is the voltage gain of the elements that is a function of the AoA, a_0 is the
signal amplitude with the ideal array gain removed,[1] and $\mathbf{x}_m$ is the position of the mth
element.[2]

The angle-of-arrival dependence of the gain of an element is the result of
foreshortening of the element area as it is viewed from an AoA direction away
from boresight and the increased impedance mismatch between the element and free
space as AoA direction increases. The former is modeled as the cosine of the angle
between the boresight direction $\hat{\mathbf{z}}$ and the AoA, ignoring any azimuthal dependance.[3]
The latter also is modeled as a cosine factor, and we taper the element gain as

$$G_e(\mathbf{k}) = |g_e(\mathbf{k})|^2 - G_{e,0}(k_z/k_0)^{\alpha_{CT}},$$

---

[1] In the link margin calculations in Chap. 2, the ideal maximum antenna gain is usually used to
compute SNR, perhaps with weighting loss included directly or included as an implementation loss.
Thus, we remove the antenna gain in the signal amplitude, and then apply a nonideal version back in
this expression. The nonideal gain, for example, includes the effects of pointing error, element
failures, and vector modulator and quantization noise.

[2] For notational compactness, we write the positions and weights of the elements as 1-D vectors. But
in a simulation and in thought, it is more convenient to represent these as 2-D arrays representing
horizontally and vertically arranged elements.

[3] There should be some dependence on azimuth for noncircular element shapes, but this is deemed
to be weak compared to the elevation angle dependence.

where $G_{e,0} = g_{e,0}^2$ is the element gain at the boresight angle, and recognizing that $\hat{\mathbf{k}} \cdot \hat{\mathbf{z}} = k_z/k_0$ is the cosine of the angle between the two unit vectors. The cosine taper exponent α_{CT} is thought to be in the range 1.4–1.6 for a well-designed AESA.

The AWGN of an element is assumed to be generated internally (i.e., not in the sky in any significant way, consistent with RF in the tens of GHz range of modern AESAs), and it too is attenuated by the element weights. Thus, we can write noise voltage as

$$v_N = \sum_{m=1}^{N_e} w_m n_m,$$

where n_m is the AWGN from the mth element. Recalling the normalization of the weights that we use and the properties of AWGN, the noise power out of the array is just

$$P_N = \sum_{m=1}^{N_e} \sum_{m'=1}^{N_e} w_m w_{m'}^* \langle n_m n_{m'}^* \rangle = P_{N,1} \sum_{m=1}^{N_e} |w_m|^2 = P_{N,1},$$

where $P_{N,1}$ is the mean noise power from an element. Without loss of generality in a simulation, we can set the noise power to unity and base the signal amplitude on the SNR.[4] Thus, the signal plus noise voltage from the array is

$$v = a_0 g_{e,0} (k_z/k_0)^{\alpha_{CT}/2} \sum_{m=1}^{N_e} w_m \exp\left[j(-\mathbf{k} + \mathbf{k}_0) \cdot \mathbf{x}_m \right] + v_N. \tag{5.2}$$

For simulation purposes, we can generate the sum and difference beams from this expression using the Taylor and Bayliss weights across the full array. Conceptually, the difference beams will be formed separately in the $x-$ and $y-$directions with full array weighting in the opposite directions. That is, for the $x-$direction monopulse measurement, the weights in the $y-$direction will be those for 32-element Taylor weighting.

In a simulation, the amplitude in Eq. (5.2) in terms of the SNR must be modified by the peak gain that is part of the SNR calculation. That is,

$$a_0 = \sqrt{\text{SNR}/G_{Rx}} \text{ (One-Way Communications Link Sum and Difference Beams).}$$

---

[4]Conversely, the noise power could be based on Eq. (2.2). and the signal power set relative to the noise floor by the SNR.

5.2   Ideal Performance of Phase-Comparison Monopulse

The ideal performance versus SNR (no failed elements or quantization or VM noise) in the $x-$ and $y-$direction (top and bottom frames, respectively) of our example AESA with uniform weighting is shown in Fig. 5.2. The average error and its standard deviation in units of the boresight beamwidth are shown in the left and right frames, respectively, of each figure row. The blue dots (pointing and AoA along boresight direction) are the same in both figures, as they should be for a symmetric beam. For the $x-$direction, the other curves are ordered N (cyan dots), NW (magenta dots), and W (black dots) from best to worst. This ordering is reversed in the $y-$direction. Although hard to see on the scale of these plots, the NW curves also are the same for the two directions. Apparently larger errors result when the measurement direction is orthogonal to the constant gain circle than occur when the measurement direction is parallel to that circle.

The average errors at low SNR values are understood by recognizing that the measurements are mostly noise and should be zero mean, and that we are pointing along boresight ($k_{0,\,x} = k_{0,\,y} = 0$). Then from Eqs. (3.2) and (3.3), the average errors

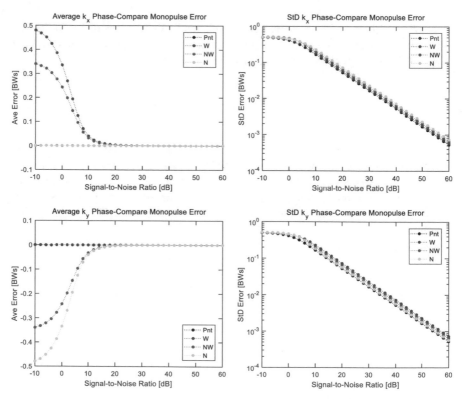

Fig. 5.2 Ideal phase-comparison monopulse in x (top row) and y (bottom row) directions with uniform weighting

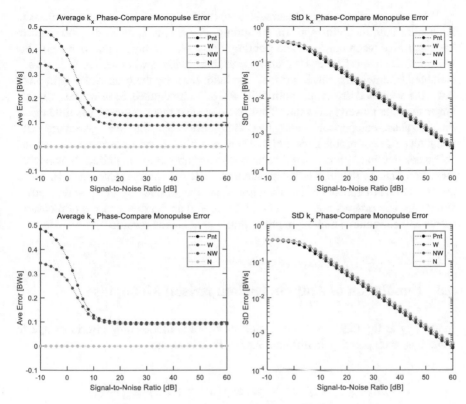

Fig. 5.3 Phase-comparison monopulse errors versus SNR with Taylor Full Array and Split-Taylor (top row) or Bayliss (bottom row) subarrays weighting

are just $+k_x$ and $-k_y$, recognizing that $-\mathbf{k}$ is the direction of arrival. For larger SNR values, the average errors all go to zero, indicating that the phase-comparison monopulse measurements are unbiased in this case.

These curves show that a factor of 10 drop in the monopulse error requires a 20-dB increase in the SNR, indicating the $1/\sqrt{\text{SNR}}$ dependence of the monopulse error, consistent with the approximate error StD reported by Sherman and Barton (2011). They give a first-order formula for the on-axis error as $\Theta_0/\sqrt{2\text{SNR}}$. At 60 dB SNR, this formula predicts a StD error of 7.07×10^{-4} beamwidths, almost exactly the same as the simulation results in the figure.

A question that arises, at least in the author's mind, is the effect of weighting on the monopulse error, and we consider Taylor weighting for the sum signal and Split-Taylor (top frames) or Bayliss (bottom frames) weighing for the difference signals in Fig. 5.3. When we look at Fig. 4.1 right frame we see that Taylor weighting reduces the peak gain from 35.1 dB to 34.2 dB and expands the beamwidth of the sum beam from $3.05°$ to $3.95°$ relative to uniform weighting. Similar changes occur on the difference beams.

When we plot the errors in beamwidths of the full aperture, these errors appear to be smaller than those for uniform weighting. For example, at 60 dB, the uniform weighted case when the AoA and pointing directions are both along boresight, the uniform StD error is 5.29×10^{-4} unweighted beamwidths, whereas it is 4.09×10^{-4} weighted beamwidths. But if we scale the latter error by the beamwidth ratio 3.95/3.05, the weighted errors are both 5.30×10^{-4} unweighted beamwidths, slightly larger than the unweighted value. Thus, we conclude that weighting has little effect on the phase-comparison monopulse StD error. But the average errors with weighting show a problematic trend. There is a bias in the measurements that the StD hides but that is a real-world issue and that is absent in the uniformly weighted case. Fortunately, this bias is independent of SNR for values greater than 10 dB or so, but it does depend on the direction (and magnitude) of the pointing error, something that often is not known beforehand, and an iterative process of measurements and corrections may be required that will consume resources.

5.3   Simulation of Amplitude-Comparison Monopulse

In analogy to Eq. (5.2), the voltages for the two amplitude-comparison monopulse subarrays, each pointing in different squint directions, are

$$v_1 = \beta a_0 g_{e,0} (k_z/k_0)^{\alpha_{CT}/2} \sum_{m=1}^{N_e/2} w_m \exp \left[j(-\mathbf{k} + \mathbf{k}_{S1}) \cdot \mathbf{x}_m \right] + v_{N1}$$

$$v_2 = \beta a_0 g_{e,0} (k_z/k_0)^{\alpha_{CT}/2} \sum_{m=N_e/2+1}^{N_e} w_m \exp \left[j(-\mathbf{k} + \mathbf{k}_{S2}) \cdot \mathbf{x}_m \right] + v_{N2},$$

where we have assumed that the numbering of the elements is such that the smaller numbered elements are in subarray "1" and larger numbered elements are in subarray "2." The scale factor β, to be determined, is there to ensure that the signal power at the output of the subarrays is correct relative to that of the full array and to the noise.

The subarray noise powers are

$$\langle v_{N1} v_{N1}^* \rangle = P_{N,e} \sum_{m=1}^{N_e/2} |w_m|^2$$

$$\langle v_{N2} v_{N2}^* \rangle = P_{N,e} \sum_{m=N_e/2+1}^{N_e} |w_m|^2,$$

where $P_{N,e}$ is the noise power out of an element. We have argued before that in order to preserve noise energy (i.e., to keep the analog noise temperature constant from element to subarray to array) it is necessary that these two be equal and that a passive power combiner be used if the two voltages are summed, which is not done in this case. Then it is necessary that

$$\sum_{m=1}^{N_e/2} |w_m|^2 = \sum_{m=N_e/2+1}^{N_e} |w_m|^2 = 1$$

that differs from the full array normalization given in Eq. (2.4).

The peak signal power out of a subarray is obtained by pointing both subarrays in the **k** direction:

$$P_{S,1} = \beta^2 a_0^2 G_{e,0}(k_z/k_0)^{\alpha_{CT}} \left| \sum_{m=1}^{N_e/2} w_m \right|^2$$

$$P_{S,2} = \beta^2 a_0^2 G_{e,0}(k_z/k_0)^{\alpha_{CT}} \left| \sum_{m=N_e/2+1}^{N_e} w_m \right|^2 .$$

Assuming that the element weights are distributed symmetrically between the two subarrays, each of the two terms must be equal to one-half the full signal power of the array. This is accomplished by setting $\beta^2 = 1/2$, and a_0^2 is obtained from the SNR of Chap. 2 but is reduced by the ideal gain used in the link margin or radar range equations. Thus,

$$\beta a_0 = \sqrt{\frac{\text{SNR}}{2G_{Rx}}} \quad \text{(One-Way Link Squint Beams)}$$

Nonideal behavior of amplitude-comparison monopulse due to element failures or quantization or VM noise is the same as that for phase-comparison monopulse.

5.4   Simulated Performance of Ideal Amplitude-Comparison Monopulse

Earlier, in Fig. 3.5, we showed the amplitude-comparison StD error versus squint angle, how to find the optimum value, and the error versus SNR at the optimum squint.[5] Now we can compare in Fig. 5.4 those results to simulation results. The solid blue lines are the analytic curves, and the blue dots are simulation results. Uniform weighting is used. The minimum error versus squint angle, in the right frame, occurs

---

[5] Appendix A has a discussion on how to convert the desired squint angle positions into pointing vectors, which is a nontrivial exercise.

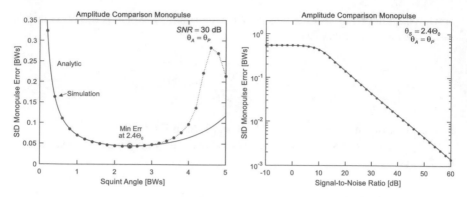

Fig. 5.4 Simulated amplitude-comparison monopulse StD error versus squint angle

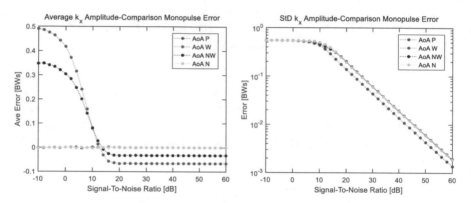

Fig. 5.5 Amplitude-comparison monopulse errors versus SNR with uniform weighting

at exactly the same value for both, and the simulation results follow the analytic results for values around or less than the optimum point. For larger squint angles, the difference between a Gaussian beam (the basis of the analytic curve) and the actual beam profile of an AESA begins to affect the comparison significantly. These results indicate that one should not try to operate amplitude-comparison monopulse with the squint angle (in full array beamwidths) greater than $3\Theta_0$ or so.

The right frame in the figure shows similarly exact agreement between the optimum error curve versus SNR. As these simulation results depend on the construction of AESA array and subarray output voltages and our careful attention to the signal and noise powers, the agreement validates the simulation in both the amplitude- and phase-comparison monopulse cases.

Amplitude-comparison monopulse performance is shown in Fig. 5.5 for the same range of SNR and for the same pointing angle errors as for the phase-comparison case. The StD errors in the right frame for nonzero pointing all fall on top of one another and slightly to the right of the ideal case of no pointing error. The StD error at 60 dB (1.36×10^{-3}), for example, is more than twice that for phase-comparison monopulse at the same SNR (5.29×10^{-4}). Furthermore, the average errors in the right frame are nonzero at large SNR values for the W and NW cases where the squint beams are not parallel to the constant gain surface. These errors represent a bias in the measurements that may be difficult to deal with in the real world.

Based on these results, we recommend phase-comparison monopulse be implemented in modern AESAs. We will not consider amplitude-comparison monopulse further, and subsequent results are for uniformly weighted arrays.

Chapter 6
One-Way Monopulse Performance in the Real World

Abstract In this chapter are descriptions of simulations of phase-comparison monopulse including real-world effects such as element failures, quantization noise, and vector modulator noise. We find some unexpected dependencies of the results on what is allowed to vary from Monte Carlo case to case. In particular, the mean monopulse error depends on whether the AWGN is correlated or uncorrelated from case to case.

Keywords Real-world monopulse performance · Element failures · Quantization noise · Vector modulator noise · Beamwidth-limited pointing · Scan angle-limited pointing

6.1 Simulation of AESA Imperfections

When we discuss quantization and vector modulator noise in the weights, we must separate what can be affected by AESA imperfections and what cannot. Ideally, for the mth element with pointing, the complex weight applied by the VM is

$$w_m \exp\left[\,j\mathbf{k}_0 \cdot \mathbf{x}_m\right].$$

The exponential term is the phase ramp applied to the weights to point the AESA in the $\mathbf{k}_0$ direction. It is this quantity that is corrupted by imperfections. The other parts of the received signal in Eq. (5.2), the plane wave and the gain of the elements in the plane-wave direction $-\mathbf{k}$, are assumed to be ideal throughout this discussion. So, for simulation purposes, we consider the received signal plus noise voltage to be

$$v_{S+N} = \underbrace{g_{e,0}(k_z/k_0)^{\alpha_{CT}/2}}_{\substack{\text{Gain of Elements} \\ \text{in Plane-Wave} \\ \text{Direction}}} \sum_{m=1}^{N_e} \underbrace{w'_m \exp\left[\,j\mathbf{k}_0 \cdot \mathbf{x}_m\right]}_{\text{Affected by AESA Imperfections}} \underbrace{a_0 \exp\left[-j\mathbf{k} \cdot \mathbf{x}_m\right]}_{\text{Rx Plane-Wave}} + v_N,$$

where the primed weight can vary from its ideal value.

© The Author(s), under exclusive license to Springer Nature Switzerland AG 2022

R. A. Dana, *Monopulse Measurement with Active Electronically Scanned Arrays (AESAs)*, https://doi.org/10.1007/978-3-030-91908-5_6

The array output noise voltage v_N can be simulated in one of two ways:

$$v_N = \begin{cases} \displaystyle\sum_{m=1}^{N_e} \underbrace{w'_m \exp[\,j\mathbf{k}_0 \cdot \mathbf{x}_m]}_{\text{Affected by AESA Imperfections}} \; v_{N,m} & \text{Modified Noise From Affected Elements} \\[3ex] \displaystyle\sum_{m=1}^{N_e} \underbrace{w_m \exp[\,j\mathbf{k}_0 \cdot \mathbf{x}_m]}_{\text{Not Affected by AESA Imperfections}} \; v_{N,m} & \text{Un-Modified Noise From Affected Elements} \end{cases},$$

where $v_{N,\,m}$ is the noise voltage from the mth element, and the primed weights are nonideal.

The idea of the unmodified case is that the noise temperature remains constant because the elements are still connected electrically and thermally to a passive power combiner. Whereas in the modified case, the noise is generated internally to the element and is affected by the changed weights. Then Eq. (2.4) can be either greater than or less than unity. In this case we rely on thermal conductivity to conserve noise energy. Both of these are idealized limits, but ones that may bound the actual much more complicated thermal situation.

Another complication in the energy balance is the possible existence of amplifiers in the summation network. These may exist to compensate for "insertion losses" in the summation network to maintain the signal level at a desired value. When this happens [e.g., see Friis equation derivation in Dana (2019), Eq. 2.16], the noise generated after the element has less effect on the overall noise temperature. Then, failed elements could indeed be modeled as outputting no signal and no noise.

Simulation of Element Failures

There are a number of ways one could model element failures, including "Off Failures" where the element output is zero (no signal, no noise), or "Signal Failures" with no signal but still with noise, or "Distortion Failures" (distorted signal, and still with noise). Because the latter can occur in any number of ways that is beyond the scope of this discussion, we assume the former mechanisms of no signal from failed elements, simulated as $w_m = 0$ for the signal, with either modified or unmodified noise.

Simulation of Quantization Noise in the Weights

The amplitude and phase are separately quantized to some number of bits. For phase, the Least-Significant-Bit (LSB) is just

$$\Delta_\varphi = 2\pi/2^{N_\varphi},$$

where N_φ is the number of phase bits. Then for a desired phase ϕ_m of the mth weight, the quantized version is

$$\phi_{m,Q} = \Delta_\varphi \lfloor \phi_m/\Delta_\varphi \rfloor,$$

where the brackets $\lfloor x \rfloor$ denote the floor function. The un-quantized phase of the element weight is computed as

$$\phi_m = \tan^{-1}\left[\frac{\text{Im}(W_m)}{\text{Re}(W_m)}\right], \tag{6.1}$$

using a four-quadrant version of the inverse tangent. The difference $\phi_m - \phi_{m,\,Q}$ is referred to as "Phase Quantization Noise."

For amplitude, we specify the Dynamic Range (DR) of the quantization and the value of the LSB (Δ_{Pw}), usually set to 0.5 dB. It is not required that the DR be a factor[1] of 6, but that seems reasonable for most cases. Then the minimum nonzero weight power is

$$P_{W,\min}[\text{dB}] = P_{W,\max}[\text{dB}] - DR[\text{dB}],$$

and the min and max power of the weights are based on the values of the weights. The quantized amplitude power [in dB] of the mth weight $P_{k,\,W}$ is

$$P_{m,Q} = P_{W,\min} + \Delta_{Pw}\left\lfloor\frac{P_{m,W} - P_{W,\min}}{\Delta_{Pw}}\right\rfloor,$$

where this expression also is in decibels. The difference $P_{m,\,W} - P_{m,\,Q}$ is referred to as "Amplitude Quantization Noise." The quantized m^{th} element weight is reconstructed as

$$w_{m,Q} = 10^{(P_{m,Q}/20)} e^{\,j\phi_{m,Q}}.$$

Simulation of Vector Modulator (VM) Noise

After quantization, the weights are transmitted from the array controller to the elements either as complex numbers or as amplitude and phase. The element, either

---

[1] Voltage is quantized. Thus, an additional bit increases the voltage range by a factor of 2 and the power range by a factor of 4 or 6 dB.

on Tx or Rx, modifies the signal, using a vector modulator, by its complex weight. But the transmission line from the controller to the element is also an analog circuit that contains white noise that corrupts the amplitude and phase commands.

We consider two ways to model VM noise, one we refer to as additive and one as multiplicative. The former, additive, is what is described by Mailloux (2005, Eq. 7.1) when he presents the distorted signal voltage, written in our notation as

$$v_S = a_0 g_{e,0} (k_z/k_0)^{\alpha_{CT}/2} \sum_{m=1}^{N_e} w_m (1 + \delta_{m,N}) \exp \left[j(-\mathbf{k} + \mathbf{k}_0) \cdot \mathbf{x}_m + j\phi_{m,N} \right], \quad (6.2)$$

where $\delta_{m,N}$ and $\phi_{m,N}$ are random amplitude and phase noise errors, respectively. These error quantities are each assumed to be normally distributed, uncorrelated, with zero mean. The Root Mean-Squared (RMS) error values σ_{Amp} and σ_{Phs} are parameters of the error analysis, and the error quantities are simulated as

$$\delta_{m,N} = \sigma_{Amp} \xi_m \quad \phi_{m,N} = \sigma_{Phs} \zeta_m,$$

where ξ_m and ζ_m are normally distributed, uncorrelated, with zero mean and unity variance.

Multiplicative noise occurs when we assume that the weight power in decibels is sent to the element, something that only would happen if there is a processor behind each element with the power to compute the weight amplitude from its power in decibels. Nonetheless, this case is of interest as an alternate model for VM noise. In this case, the mth weight power in decibels is

$$W_{dB} = 20 \log_{10}(|w_m|).$$

To this, noise is added in the control line to the element, and the weight arriving at the element is

$$W'_{dB} = W_{dB} + \sigma_{dB} \nu_m,$$

where ν_m is normally distributed, zero-mean, and unity variance, uncorrelated from the phase error term. Because we want to compute the weight voltage using an exponential function since the phase error is just $e^{j\varphi_m}$, we convert the weight power in decibels to an exponent:

$$w' = \exp \left(\alpha_{dB} W'_{dB} + j\phi_m + j\phi_{m,N} \right),$$

where again ϕ_m is the phase of the weight without noise in Eq. (6.1), and the constant α_{dB} converts a voltage quantity expressed in decibels to an exponent:

$$\exp\left(\alpha_{dB}x\right) = 10^{(x_{dB}/20)} \quad \alpha_{dB} = [20\log_{10}(e)]^{-1}.$$

Inserting the noise corrupted weight power into the w' expression we find that

$$w'_m = |w_m| \exp\left(\, j\phi_m\right) \, \exp\left(\alpha_{dB}\sigma_{dB}\nu_m + j\phi_{m,N}\right) = w_m \exp\left(\alpha_{dB}\sigma_{dB}\nu_m + j\phi_{m,N}\right).$$

In further developments, we will vary these AESA noise sources parametrically to observe the effects on important parameters such as peak gain and monopulse measurement error.

6.2   Real-World Limitations to Phase-Comparison Monopulse Performance

An AESA deployed in the real world will have imperfections, but it also has situations that stress the beamwidth and pointing range of the array. These are discussed first, followed by the effects of array imperfections: element failures, and quantization and vector modulator noise.

Limitations of AESA Beamwidth and Pointing Range

Often it is the case, particularly when a new signal is being acquired, that the array pointing direction is not closely aligned with the true AoA. The question is then by how much can the two differ and still have a monopulse measurement that will allow the next observation to be closer to the true AoA. Assuming boresight pointing, the mean and StD errors are plotted in Fig. 6.1. The AoA is increasing along the k_x axis in this example. As will be the case with a number of the results in this chapter, the SNR for this plot is set at 30 dB when the pointing direction is aligned with the AoA and both are at boresight, which is the zero point of the abscissa.

First consider the right frame where the StD error is plotted for the $x-$direction (blue curve) and the $y-$direction (magenta curve). As the array is uniformly weighted in this case, the first null in the beam pattern, measured in 3-dB FWHM beamwidths, occurs at 1.1288 on the abscissa (see Appendix B), exactly where the StD errors are largest (our 0.1 stepping of the simulation points misses this null by a small amount). Beyond the null, the phase difference between the two subarrays becomes ambiguous,[2] which accounts for the large increase in the average k_x error (right frame) after crossing the null. In the $y-$direction, the boresight beam and AoA

---

[2] If the AoA of a plane wave is within the main beam, null to null, then the phase difference of the plane wave from one side of the array to the other is less than a wavelength. Outside of this angular extent, the phase difference from side to side of the array is ambiguous (i.e., greater than 2π).

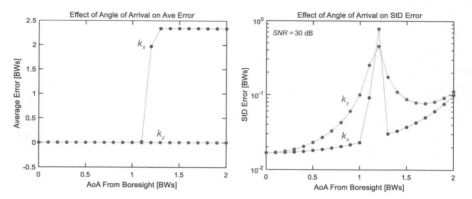

Fig. 6.1 Phase-comparison monopulse measurement mean (left frame) and StD (right frame) error as function of AoA from boresight pointing

are aligned, and the mean error remains near zero as the phase difference in this direction is not ambiguous.

We conclude that one should ignore AoA's measured with phase-comparison monopulse more than about one beamwidth from the pointing direction. Even though the StD error may be reasonable in the plot beyond this point, the measurements know nothing about the mean error (or an error bias), and this cannot be corrected in this case without knowing the truth (a distinct advantage of a simulation versus life in the real world).

Limitations of AESA Scan Angle

One great advantage of a phased array is that it can point its beam away from boresight without having to mechanically rotate the array; so, no moving parts, no complicated pedestal wiring or waveguides, no missed airline departures because of failed mechanics. But there is a limit to how far away one can electronically scan a beam and still have sufficient received power to make monopulse measurements, recalling that because of the cosine taper of an element it has zero gain parallel to the array face. Phase-comparison monopulse error curves (in beamwidths measured at boresight) for this case are shown in Figs. 6.2 and 6.3, where the abscissa is the elevation angle of the scan, which is done along an azimuth of $45°$, in which case the k_x and k_y StD errors are the same (and fall on top of one another in the right plot). These results indicate that the StD error increases rapidly for scan angles beyond $60°$ or so.

What is surprising about these results is that the mean errors in the two directions are different at boresight (zero elevation angle) and have the different trajectories. Note that the ordinate range of the left plot is still small compared to the corresponding StD errors. This odd behavior is a consequence of an initial simulation development choice that reset the AWGN random number generator to the same point for each elevation angle, i.e., the AWGN samples are identical for each

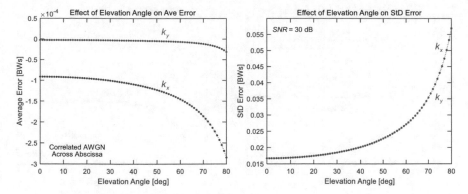

Fig. 6.2 Phase-comparison monopulse measurement error mean (left frame) and StD (right frame) as function of scan angle from boresight with correlated AWGN

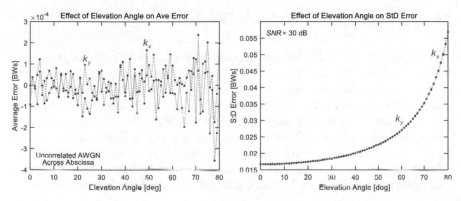

Fig. 6.3 Phase-comparison monopulse measurement mean (left frame) and StD (right frame) error as function of scan angle from boresight with uncorrelated AWGN

elevation angle simulation. Then the question is: is this valid or not, and what happens if the AWGN sequences are different for each elevation angle? Because an AESA cannot scan to two elevation angles simultaneously using the same elements, the measurement at different scan angles in the real world is sequential or must use different subarrays for each scan angle. Either way, the AWGN from scan angle to angle should be uncorrelated, and results for this situation are shown in Fig. 6.3. Now the mean errors are randomly uncorrelated between the two directions and from one scan angle to the next. But because the mean errors are small compared to the StD errors in either case, the StD errors are essentially the same for both.

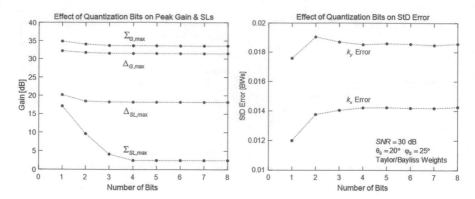

Fig. 6.4 Effect of weight quantization on array gain and sidelobe level (left frame) and phase-comparison monopulse measurement error StD (right frame) versus number of bits

Limitations of Number of Bits in Weights

The weights are sent to the elements on analog lines that carry a limited number of bits for the amplitude and phase. In Fig. 6.4 are plots of the peak gain and maximum sidelobe levels for the sum and $x-$direction difference beam versus the number of bits, kept the same for both amplitude and phase. Taylor/Bayliss weights are used, and the beam is pointed at the target at $25°$ azimuth and $20°$ elevation. The monopulse measurement StD errors are plotted in the right frame for an SNR of 30 dB. Oddly, these errors vary slightly as the number of bits is increased, but the primary effect of quantization bits is in the peak sidelobe levels that for the sum beam does not reach its intended value until the number of bits is at least 4.

Limitations of Failed Elements

Often an AESA is an integrated structure, and elements cannot be replaced singularly as they fail. Failed elements will degrade phase-comparison monopulse performance as they affect the true subarray phase center location. Of course, other metrics degrade as elements fail, such as peak gain, maximum sidelobe level, integrated sidelobe level, and beamwidth.[3]

As mentioned above, we simulate hard failures of an element by completely turning off the output signal ($w_m = 0$ for the signal), but the noise can be either off also or worst case can remain unmodified from its original statistics. The AESA SNR gain and the k_x phase-comparison monopulse StD errors are plotted in Fig. 6.5 for the

---

[3] Unless the elements fail in a systematic manner, e.g., an entire element row or column at the edges of the array, element failures have little effect of the beamwidth that is primarily a function of array extent. We simulate the failed elements as uniformly distributed in the two dimensions of the array.

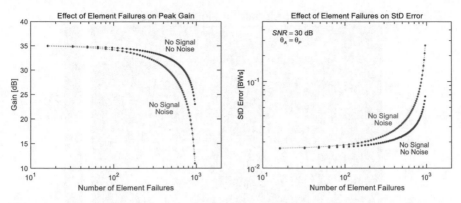

Fig. 6.5 AESA gain (left frame) and phase-comparison monopulse StD error (right frame) versus number of element failures

"No Signal, No Noise" case (blue curves) and the "No Signal, Noise" case (magenta curves). As one would expect, the no signal, no noise case has higher gain and lower StD error compared to the other case. We expect that reality falls somewhere between these two sets of curves. As the "No Signal, Noise" case represents a reasonable worst case, we will use it in further results.

What these results show is that monopulse error and array gain are remarkably insensitive to element failures as long as the fraction of failures remains less than say 20 percent. Above this level, both the gain and monopulse error rapidly degrade.

Limitations of Vector Modulator (VM) Noise

The weights are transmitted to the elements via analog control lines that themselves have noise that corrupts the received weights. This noise can be either additive or multiplicative, depending on whether the actual weight amplitude and phase are transmitted (additive) or the amplitude represented by its decibel power (multiplicative). We simulate both cases, as both are possible depending on the AESA design.

Figure 6.6 shows contours of the peak gain for a return with AoA and array pointing direction along boresight. Ideally, this gain is $10\log_{10}(1024\pi) = 35.07$ dB. The abscissa is the RMS amplitude error (these errors have zero mean), and the ordinate is the RMS phase error. These VM errors are additive. The range of the abscissa is both negative (RMS errors are less than unity) and positive (RMS errors are greater than unity). That the contours are mostly vertical indicates little sensitivity of the peak gain to phase errors over its range ($0° - 20°$).

Figure 6.7 shows contours of the corresponding phase-comparison monopulse mean (left frame) and StD (right frame) errors in the $x-$direction for the same VM errors. The error plots for the $y-$direction are very similar. That the StD contours are mostly vertical, the monopulse errors are insensitive to phase errors over its range.

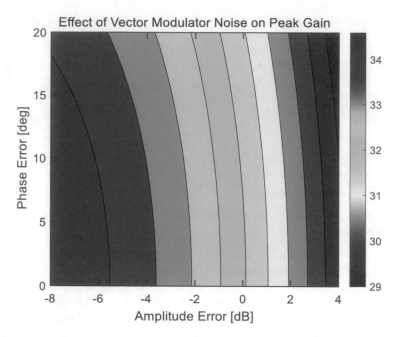

Fig. 6.6 Additive VM error peak gain of array (along AoA) versus amplitude (abscissa) and phase (ordinate) RMS errors in the VM weight commands

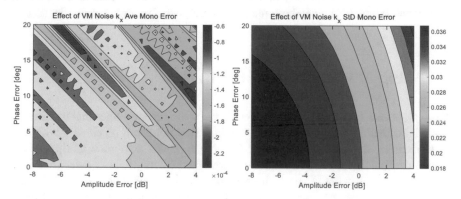

Fig. 6.7 Additive VM mean (left frame) and StD (right frame) phase-comparison monopulse errors versus amplitude and phase errors in VM weight commands

Also, note that the mean error is much smaller than the StD, and the latter varies by a factor of 2 or less for these VM error ranges.

Similar contour plots for multiplicative VM errors are shown in Figs. 6.8 and 6.9. In this case, VM RMS amplitude error varies from 0 (no amplitude error) to 3 dB (factor of two multiplicative RMS amplitude error). These contour plots differ from the additive ones because in the former the amplitude RMS error is never zero as it is along the left axis in this case.

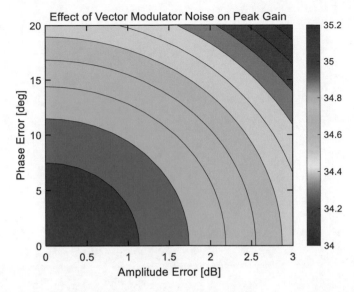

Fig. 6.8 Multiplicative VM error peak gain of array (along AoA) versus amplitude (abscissa) and phase (ordinate) RMS errors in the VM weight commands

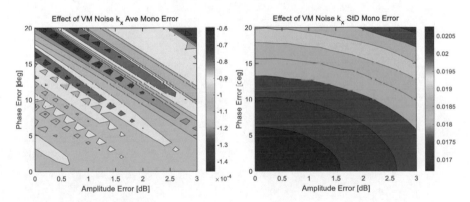

Fig. 6.9 Multiplicative VM mean (left frame) and StD (right frame) phase-comparison monopulse errors versus amplitude and phase errors in VM weight commands

For multiplicative VM errors, the range of peak gain (about 1 dB) is smaller than that for additive errors (about 5 dB), but keep in mind that there is not a 1 to 1 correspondence in the RMS amplitude errors between the two cases. Again, the RMS monopulse error range is small, a factor of two or less in both cases. We conclude that VM noise, at least for the ranges we simulated, has less of an effect on monopulse performance than do the other error sources we investigated.

Chapter 7
Two-Way (Monostatic Radar) Monopulse Performance

Abstract A monostatic radar, when performing monopulse measurements on a target, transmits the signal toward the target using the full aperture (sum beam). On return, that signal is received by subarrays from which the sum and difference beams are formed. Subsequent phase-comparison monopulse measurements are exactly the same as those for the one-way case. A primary difference between the one- and two-way cases is that the sum beam gain enters the problem twice, once on transmit and once on receive.

Keywords Radar monopulse · Monostatic radar · Radar cross section (RCS) · Swerling RCS models · RCS samples · Probability of false alarm · Probability of detection · Marcum's Q-function

7.1 Two-Way Gain Computed Step by Step

The best way to understand the contributions to the monostatic radar received signal from a radar target is to construct the signal step by step, i.e., start with the transmitted signal in space and follow it to and from the target and into the radar signal processor. To that end, consider a sum beam transmitted at a target in the direction $\mathbf{k}$ when the sum beam is pointed in the direction $\mathbf{k}_0$, not too far from the true direction of the target. To simplify this calculation, we consider the antenna to be a continuous aperture, not the sum of discrete elements. The voltage gain of the Tx antenna pointed at the target is

$$g_{Tx}(\mathbf{k}) = \int\limits_{-\infty}^{\infty} w_{Tx}(\mathbf{r}) \exp\left[\, j(\mathbf{k} - \mathbf{k}_0) \cdot \mathbf{r}\right] d^2\mathbf{r},$$

where $\mathbf{r}$ is a vector in the plane of the aperture, $w_{Tx}(\mathbf{r})$ is the transmit weighting applied to the aperture that is zero where the aperture is not, and the 2-D integral is over the infinite extent of the aperture plane for mathematical convenience. The sign of the exponent is opposite that from before because this expression is for a Tx gain, not the Rx gain, for example, in the Eq. (5.2) expression.

At the target at vector range $\mathbf{R}$ from the phase center of the Tx aperture, the signal (in voltage units) incident on the target is

$$v_T(\mathbf{R}) = \frac{\sqrt{P_{Tx}}g_{Tz}(\mathbf{k})\exp\left(\,jk\cdot\mathbf{R}\right)}{\sqrt{4\pi R^2}} \quad \left[\sqrt{W/m^2}\,\right],$$

where $R = \|\mathbf{R}\|$ is the slant distance to the target. Because the direction of propagation must be toward the target to get a return, $\mathbf{k}\cdot\mathbf{R} = 2\pi R/\lambda$, even though the AESA may not be pointed exactly in that direction. The units of $|v_T(\mathbf{R})|^2$ are power per unit area.

The target intercepts a portion of the signal that is reflected back toward the radar; its effective scattering area is described by its RCS σ that has units of area. So, the signal reflected by the target RCS in the direction of the radar is

$$v_\sigma(\mathbf{R}) = v_T(\mathbf{R})\sqrt{\sigma}\exp\left(\,j\phi_\sigma\right) = \frac{\sqrt{P_{Tx}}g_{Tz}(\mathbf{k})\exp\left(\,jk\cdot\mathbf{R}+j\phi_\sigma\right)}{\sqrt{4\pi R^2}} \quad \left[\sqrt{W}\,\right],$$

where ϕ_σ is the phase shift associated with the voltage RCS.

After propagating the distance R back to the radar, the signal incident on the receive aperture at position $\mathbf{r}$ in the Rx aperture plane is

$$v_R(\mathbf{r}) = \frac{v_\sigma(\mathbf{R}+\mathbf{r})\exp\left[\,jk\cdot(\mathbf{R}+\mathbf{r})\right]}{\sqrt{4\pi R^2}}.$$

Integrating this signal over the area of the Rx aperture, we get the following first-principles expression for the received voltage, assuming that $\|\mathbf{R}+\mathbf{r}\|\approx R$ over the surface of that aperture:

$$\boxed{v_{Rx}(\mathbf{r}) = \frac{\sqrt{P_{Tx}\sigma}}{4\pi R^2}g_{Tx}(\mathbf{k})\exp\left(2jk\cdot\mathbf{R}+j\phi_o\right)\int_{-\infty}^{\infty}w_{Rx}(\mathbf{r})\exp\left[-j(\mathbf{k}-\mathbf{k}_0)\cdot\mathbf{r}\right]d^2\mathbf{r}.}$$

The integral should be recognized as the received voltage gain, so the received voltage out of an Rx beam is

$$v_{Rx}(\mathbf{r}) = a_0 g_{Tx}(\mathbf{k})g_{Rx}(-\mathbf{k})\exp\left(\,j4\pi R/\lambda+j\varphi_\sigma\right). \tag{7.1}$$

The product of the two gains in k-space is equal to the convolution of the Tx and Rx weight function in physical space. This equation expresses the R^{-4} dependence of the power of a monostatic radar signal, and its $4\pi R/\lambda$ phase shift due to propagation distance. The Rx gain can be either for the sum beam or the difference beams.

Noting that monopulse difference beams are formed only on receive and the full aperture sum beam is used on transmit, the relationship between the amplitude in Eq. (7.1) and SNR is

$$a_0 = \sqrt{\frac{SNR}{G_{Tx}G_{Rx}}} \frac{\sigma}{\langle \sigma \rangle} \quad \text{(Monostatic Radar Sum and Difference Beams)}.$$

where $\sigma/\langle \sigma \rangle$ is the ratio of the RCS to its ensemble mean value that depends on the model used for RCS fluctuations.

7.2   Simulation of Monostatic Radar Phase-Comparison Monopulse

In a simulation of monostatic radar monopulse, the voltage gain integrals are replaced by sums, so the gains in Eq. (7.1) are evaluated as

$$g_{Tx}(\mathbf{k}) = \sqrt{G_e} \sum_{m=1}^{N_e} w_{m,Tx} \exp\left[j(\mathbf{k} - \mathbf{k}_0) \cdot \mathbf{x}_m \right]$$

$$g_{Rx}(-\mathbf{k}) = \sqrt{G_e} \sum_{m=1}^{N_e} w_{m,Rx} \exp\left[j(-\mathbf{k} + \mathbf{k}_0) \cdot \mathbf{x}_m \right].$$

Otherwise, the phase-comparison monopulse processing is exactly as before for one-way links.

There are standard models for RCS fluctuations that are referred to as Swerling 0, Swerling 1, and Swerling 3, named after Peter Swerling[1] (Swerling 1960) who invented them. These models are defined by their probability density functions (PDFs), $f(\sigma)$:

$$f(\sigma) = \begin{cases} \delta(\sigma/\langle \sigma \rangle - 1) & \text{Swerling 0, Non-Fluctuating} \\ \dfrac{1}{\langle \sigma \rangle} \exp\left(-\sigma/\langle \sigma \rangle\right) & \text{Swerling 1} \\ \dfrac{4\sigma}{\langle \sigma \rangle^2} \exp\left(-2\sigma/\langle \sigma \rangle\right) & \text{Swerling 3} \end{cases} \quad (\sigma \geq 0),$$

where $\delta(x)$ is the Dirac delta-function. Sometimes the Swerling 1 model is referred to as chi-square of degree 2, and the Swerling 3 model as chi-square of degree 4, a distinction that is useful in the generation of RCS samples.

---

[1] There are also Swerling 2 and 4 models that apply when there is noncoherent integration. But such integration wipes out the phase of the signal, rendering it useless for phase-comparison monopulse. Coherent integration, on the other hand, requires phase coherence of the signal over the period of the integration. In this case, the SNR is increased with the number of pulses integrated, and single-pulse expressions can be used for both target detection and phase-comparison monopulse errors.

These designations allow a straightforward way to generate random samples of the RCS for the fluctuating models. First, the cumulative distribution of the exponential PDF is invertible:

$$F(\sigma) = \int_0^\sigma f(\sigma')d\sigma' = 1 - \exp\left(-\sigma/\langle\sigma\rangle\right).$$

Setting $F(\sigma) = 1 - u$ where u is uniformly distributed $[0, 1)$ gives

$$\frac{\sigma}{\langle\sigma\rangle} = -\ln(u) \text{ (Swerling 1)}.$$

Noting that a chi-square random variable of degree 4 is the sum of two chi-square of degree 2 random variables, gives for the Swerling 3 case

$$\frac{\sigma}{\langle\sigma\rangle} = -\frac{\ln(u_1) + \ln(u_2)}{2} \text{ (Swerling 3)},$$

where u_1 and u_2 are both uniformly distributed $(0, 1]$ and are independent. The factor of ½ is there because of the mean value of $-\ln(u)$ is unity. We assume that the phases associated with the RCS are uniformly distributed and independent of the RCS amplitude for all cases.

Although easy to derive from the statistics of the noise and that of the RCS models, radar probability of detection is beyond the scope of this book and is readily available in the radar literature. So, we report results from Swerling (1960). These will be used to validate the simulation of phase-comparison monopulse for monostatic radar. The reason that probability of detection comes into play is that usually a monopulse measurement will be performed only after the detection of a target, eliminating low SNR values from the error statistics. Target detection starts with thresholding voltage samples that contain noise alone to minimize the number of samples that must be validated as being from real targets.

If the real and imaginary parts $(x + jy)$ of the noise are zero-mean, normally distributed with variance σ_N^2, then it is easy to see that the probability density function of the power of a noise sample $(P = x^2 + y^2)$ is exponential:

$$f_N(P) = \frac{1}{2\sigma_N^2} \exp\left(-P/2\sigma_N^2\right),$$

where from before $2\sigma_N^2$ is the mean noise power of the sample. The probability of false alarm is the probability that noise alone will exceed the threshold, or

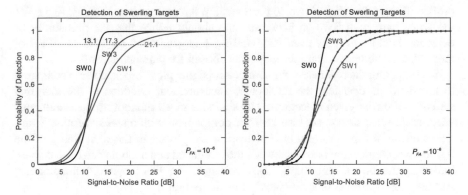

Fig. 7.1 Probability of detection of Swerling 0, 1, and 3 RCS targets

$$P_{FA} = \int\limits_{t}^{\infty} f_N(P)dP = \exp\left(-t/2\sigma_N^2\right) = e^{-T},$$

where

$$T = t/2\sigma_N^2 = -\ln\left(P_{FA}\right)$$

is the threshold-to-noise ratio. A commonly used value for the probability of false alarm is $P_{FA} = 10^{-6}$, which requires a threshold-to-noise value of 11.4 dB. Against this threshold, the probabilities of detection of the three RCS models are

$$P_D = \begin{cases} \int\limits_{t}^{\infty} f_{S+N}(P)dP & \text{General Expression} \\ Q\left(\sqrt{2S}, \sqrt{-2\ln(P_{FA})}\right) & \text{Swerling 0, Non-Fluctuating RCS} \\ P_{FA}^{\frac{1}{1+S}} & \text{Swerling 1} \\ P_{FA}^{\frac{2}{2+S}}\left[1 - \frac{2S\ln(P_{FA})}{(2+S)^2}\right] & \text{Swerling 3} \end{cases},$$

where $f_{S+N}(P)$ is the probability density function of the signal plus noise voltage power, S is the signal-to-noise ratio of the sample (not in decibels), and $Q(x, y)$ is Marcum's Q function (Marcum 1948) that results from the Rician distribution of signal plus noise voltage amplitude.

These "single dwell" (i.e., single pulse or multiple pulses coherently integrated during a dwell on target) detection probabilities are plotted versus SNR in Fig. 7.1. In the left frame, the dashed line is at $P_D = 0.9$, a reasonable goal for a tracking radar, and the SNR values required to achieve this goal for each model are indicated. For probabilities of detection above 50 percent,[2] SW0 represents the best case and SW1 the worst case with SW3 in between. One should notice that the probability of detection is close to unity even for Swerling 1 targets when the SNR is 30 dB or larger that is the reason that this value was chosen for parametric results.

In simulating the radar case, the sum beam (signal plus noise) power is compared to threshold. If detected, the monopulse measurement is added to the marginal statistics of errors given detections. The statistics of all monopulse measurements, whether detected or not, are also kept for comparison with one-way results. Simulation results for the probability of detection are plotted in the right frame of the figure. Exact agreement is seen between the simulated and analytic detection curves, indicating that both the signal plus noise voltages, Swerling target RCS fluctuations, and the detection process are correct in the simulation.[3]

7.3   Ideal Monostatic Radar Phase-Comparison Monopulse Performance

Ideal one-way (black dots), two-way Swerling 0 (blue dots), two-way Swerling 1 (magenta dots), and two-way Swerling 3 (cyan dots) phase-comparison monopulse performance is shown in Fig. 7.2. The average errors are plotted in the left frame, and the StD errors in the right frame. The average errors show that no bias is introduced in radar monopulse.

The one-way curve, for which there is no thresholding, is the same as results presented earlier. The two-way Swerling 0 curve coincides with the 1-way curve for SNR values above 15 dB where the probability of detection is nearly unity. Thus, for the same SNR in the voltage samples, one- and two-way monopulse perform

---

[2] Search and track radars are designed to achieve a large probability of detection for the RCS of interest at the maximum operating range. The process of acquisition is often as follows (e.g., Dana and Moraitis 1981): A new detection is followed quickly by a verification look at the target at the same frequency to ensure the first detection was not just noise and that the RCS does not fluctuate to a smaller value. Then after two consecutive detections, a track is started on the new target. Once in track, the target is revisited often enough to maintain the track even if the target is maneuvering by using waveforms with high-accuracy range and angular (monopulse) measurements. An AESA is ideally suited to this process as it can scan back to the target as necessary while still maintaining its search function. The reason for the high P_D and a low P_{FA} is that noise and missed detections waste radar resources, and too many false targets entering into track can overwhelm track association algorithms (i.e., a target detection during search is incorrectly associated with another target in track and not with its own track, if already established).

[3] When first plotted during code validation, the simulation detection results were about 5 dB to the right of the analytic curves. After considerable consternation, the author finally remembered that 5 dB is a "magic number" for antennas as it is the gain of an element when sized at ½ the wavelength. Sure enough, the element gain was put into the two-way gains only once rather than twice for the Tx and Rx beams. This catch is an example of why it is a good idea whenever possible to compare simulation results to known results.

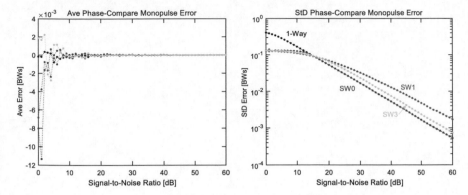

Fig. 7.2 Simulated phase-comparison monopulse errors for 1- and 2-way geometries and Swerling targets

identically when there are no RCS fluctuations in the measurements, something that is expected. However, with RCS fluctuations the performance degrades at an StD error of 3×10^{-3} by about 3 dB for Swerling 3 and by 10 dB for Swerling 1. In the latter case, the slope of the StD curve is slower than $1/\sqrt{\text{SNR}}$, indicating that this performance loss gets larger for larger SNR values.

Chapter 8
Other Real-World Effects on Radar Detection Range

Abstract Once monopulse performance has been established as a function of SNR, often the next question is something like "how far can we see?" This is particularly important for radars operating, for example, in the GHz range. Of course, one can pick an SNR value and invert the radar range equation to find the ideal maximum range. The problem is the atmosphere and water therein. In this frequency range the attenuation caused by dry air, water vapor, and rain can be severe. One should not get too carried away with fidelity in these calculations until it is clear that one has only a small margin or none at all to meet requirements. Then effects such as a non-flat Earth and refraction may play a significant role. This chapter presents an example of how to approach these real-world problems using information available in the public domain.

Keywords Power amplifiers · Analog devices HMC635 · Specific attenuation of atmosphere · Specific attenuation of rain · US standard atmosphere · International Telecommunications Union (ITU) Specific Attenuation Models · Swerling 1 target

Consider a radar operating on an aircraft usually flying at an altitude of 2 km, and you are asked how far can one see a Swerling 1 target with a 0 dBsm RCS [decibels relative to 1 square meter] and with a probability of detection of 90 percent. The requires an SNR of 21.1 dB, as shown in Fig. 7.1. You are told that the radar and AESA are just one component of a Multifunction RF (MFRF) system that must operate in the 15–40 GHz range, but will be optimized at just one frequency in that range (i.e., the element area will be $\lambda^2/4$ at the optimization frequency). Which frequency is best from the radar perspective?

A number of factors must be considered, including the AESA Power Amplifiers (PAs) at each element of the AESA. An RF expert suggests that Analog Devices part HMC635 is a good candidate because of its relatively flat response across the desired bandwidth (specified as 18–42 GHz). The HMC635 data sheet (Analog Devices 2021) has a plot of the saturated power output, appropriate for a pulsed radar, versus frequency that is captured in Fig. 8.1 (blue dots). The red dashed line is a 5th order polynomial fit to the data that allows evaluation at any frequency in the bandwidth

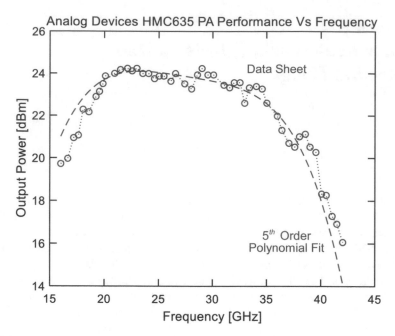

Fig. 8.1 Analog Devices HMC635 PA radiated power versus frequency

and extrapolation to frequencies above and below the bandwidth.[1] Applying this to the 1024 elements in the example AESA, the radiated power varies between 16 and 24 W, not nearly the radiated power of many ground-based radars but realistic for an MFRF device.

In order to evaluate the radar range equation, Eq. (2.3), a number of additional parameters are specified in Table 8.1. These example values are not intended to represent any particular system.

The number of pulses coherently integrated is quite arbitrary, of course, and is intended to provide enough energy on target to achieve detection at a range over 20 km, which sounds impressive for a radar not as bright as a 25 W lightbulb until one applies attenuation from the atmosphere and rain. This total energy could have been obtained by an uncompressed pulse 1.28 ms long and then compressed by a factor of 1000, or by a Frequency-Modulated, Continuous-Wave (FMCW) chirp waveform with a time-bandwidth product of 1000 and a chirp duration of 1.28 ms. It is the energy (power × time) on target that matters, not how that is achieved.

The first environmental hurdle for radars operating in this band is the effect of air attenuation that is plotted in Fig. 8.2 for links propagating at a constant 2 km altitude. This attenuation is computed by first using the U.S. Standard Atmosphere (NOAA 1976) to find the pressure and temperature at altitude, and then using the formula in ITU (2007) to compute the specific attenuation (per km) versus frequency. In the left frame are plots of dry air (red curve) attenuation (primarily from O_2), that from water

---

[1] Proof that the author is not an RF engineer.

Table 8.1 MFRF radar and AESA parameters

Parameter	Value
Altitude	2 km
AESA elements	1024
Element radiated power	16–24 dBm
Total radiated power	16–24 dBW
AESA optimization frequency	15 or 30 GHz
Required P_D of Swerling 1 target	0.9
Reference RCS	0 dBsm
Required P_{FA}	10^{-6}
Required SNR	21.14 dB
Noise figure	5 dB
Pulsewidth	10 μs
Pulses coherently integrated	128
Total time on target	1.28 ms

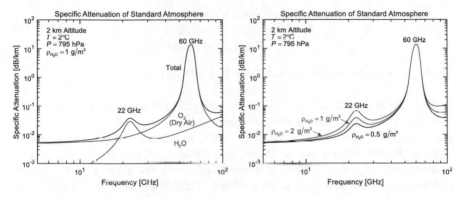

Fig. 8.2 Atmospheric specific attenuation in the 5–100 GHz band

vapor (blue curve) with 1 g/m^3 concentration, and the sum of the two (black curve). Water vapor absorption at 22 GHz and that from oxygen at 60 GHz are prominent features of these plots. The right frame shows the effect of varying water vapor for three concentrations, and only the total attenuation is plotted. The reason for considering an MFRF band from 10 to 40 GHz is clearer from these data—open bands exist below 22 GHz and from 30 to 40 GHz. And if one wants to communicate with a nearby platform, 22 GHz provides some natural Low-Probability of Intercept (LPI) protection from more distant ears.

Because water vapor is a key contributor to atmospheric attenuation, one must consider the effect of rain also. This attenuation is plotted (blue curve) in Fig. 8.3 along with wet air attenuation for three concentrations (magenta curves) for comparison. The rain data were generated using the procedure outlined in ITU (1999). The wet air attenuation comes from 1 percent, median, and 99 percent water vapor concentrations given in NOAA (1976, Table 20) (i.e., 99 percent of rain events will produce attenuation less than the plotted value whereas 99 percent will produce attenuation greater than the 1% curve).

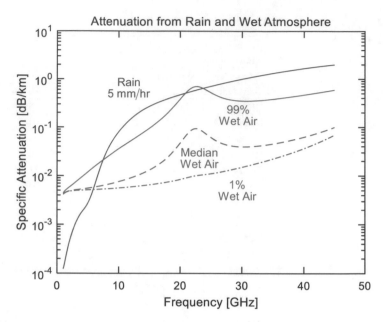

Fig. 8.3 Rain and wet air specific attenuation in the 1–50 GHz band

Even moderate rainfall rates (5 mm/h or greater) are a bigger threat to MFRF operation in this band than is wet air attenuation, as none of the curves plotted in Fig. 8.2 exceed 0.1 dB/km over the band of interest. In rain these data also indicate that one should concentrate on frequencies below 20 GHz. A key assumption, however, in using the rain attenuation to compute detection range is that it applies over the entire path from Tx to target and back, which is a worst case that may never be quite true.

Applying these attenuation values and the radar range equation, the detection ranges of a Swerling 1 target (for $P_D = 0.9$ and $P_{FA} = 10^{-6}$) are plotted in Fig. 8.4 versus frequency. The AESA is optimized at 20 GHz in the left frame and at 30 GHz in the right frame. The AESA gain at the optimum frequency is 1024π for both cases. Without any atmospheric attenuation, the detection range for 20 GHz is about 21 km, whereas it is about 17 km for 30 GHz, indicating that the weak PA frequency dependence is not enough to overcome the remaining λ^2 term in the radar range equation.

Now we are in a position to answer the program manager's primary question— how far can this radar see? "Well, it depends ..." is not an answer that program managers want to hear. But it is clear that optimizing the element size at 20 GHz provides about 1.2 times the detection range as does optimization at 30 GHz.[2] The detection range regardless of element size depends strongly on the presence of rain

---

[2]But keep in mind that we have kept the same number of elements (1024) so the 30 GHz elements occupy 0.44 of the area, and that may be a distinct advantage for an airborne radar.

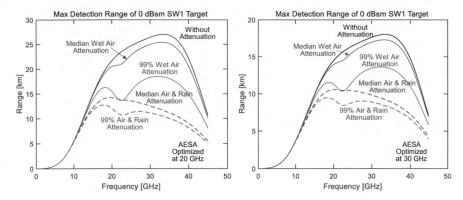

Fig. 8.4 Swerling 1 target with 0 dBsm RCS detection range versus frequency

or not. Without rain, the 20 GHz radar can see to about 15 km depending on the exact band and the humidity, whereas the smaller 30 GHz radar can see to 10 km depending on band and humidity.

If we were to keep the AESA the same size in going from 20 to 30 GHz optimization by increasing the number of elements by a factor of 2.25 and increasing the peak two-way gain by about 7 dB, then the R^{-4} dependence of the SNR says that the detection range without attenuation would increase by a factor of 1.2. So, this could give the advantage to the 30 GHz array. Keep in mind, however, that the power needed to drive the element PAs would increase by a factor of 2.25 also; something that may not be well received by program management.

Chapter 9
Examples of the Use of Monopulse Measurements in Radar Target Tracking with Kalman Filters

Abstract We have been focusing on making monopulse measurements and their accuracy. But in the real world monopulse will only be implemented if necessary to meet system requirements. An example of such use is the tracking of a radar target using a Kalman filter. However, this is a subject with a vast history in the literature, so we limit the scope of this example to tracking one target that ignores the issues of track initiation and association, but does include the effects of missed detections during track. Another important consideration that we simplify but do not ignore is that that monopulse measurements are made in the reference frame of the antenna, whereas it is usual to track targets in a navigation frame such as East-North-Up (ENU) with the AESA radar at the origin. An outline of the derivation of Kalman filters is given in Appendix C.

Keywords Kalman filters · Extended Kalman filters · Weight least-squares filters · Covariance collapse · Singer dynamics model

Consider an area defense radar based on our example AESA looking for targets intruding into a given space. Such a system might be based around high-value facilities such as dams or power plants or military bases. Another possible example is a system looking for quadcopters trying to drop contraband into a prison. This chapter presents an example Extended Kalman Filter (EKF) tracking of a target using either slant range or slant range and 2-D monopulse measurements as inputs to the tracking filter. The filter is extended because of the nonlinear relationship between the measurements and the track position.

The goal of such a filter is to present to the user with high accuracy 3-D information on the target location once it has been detected and placed into track. To this end we can ignore a number of effects with which real radar must contend. These include

- Oblate spheroid geometry of the Earth (a flat Earth is assumed).
- Atmospheric attenuation.

- Multiple targets in the search and track volume, including clutter.
- Target acquisition and track association (associate target detections with existing tracks).

9.1 Geometry and Coordinate Transformations

Monopulse measurements are made in the antenna frame that we have defined as having an x-axis along one dimension of the array face and the z-axis pointed in the boresight direction. The y-axis completes the right-handed coordinate system. But target tracking is usually done in a locally-level frame such as ENU with the AESA radar at the origin. We assume for this example that the target is tracking in an ENU system, with the AESA face sitting normal to the reference plane (a plane tangent to the Earth's surface) with its x-axis pointed up. Then the translation of a position vector $\mathbf{r}^A$ from the Antenna Frame (denoted by a superscript A) to the ENU Navigation Frame (denoted by a superscript N) of a vector pointing at the target is

$$\mathbf{r}^N = C_A^N \mathbf{r}^A,$$

where

$$C_A^N = \begin{bmatrix} 0 & 1 & 0 \\ 0 & 0 & 1 \\ 1 & 0 & 0 \end{bmatrix}.$$

The matrix that transforms a vector from the N-frame to the A-frame is just the transpose of this unitary matrix:

$$C_N^A = \left(C_A^N\right)^T = \begin{bmatrix} 0 & 0 & 1 \\ 1 & 0 & 0 \\ 0 & 1 & 0 \end{bmatrix}.$$

The more general form of these equations when the AESA has a non-orthogonal relationship to the ENU frame is discussed, for example, in Dana (2019).

9.2 EKF Matrices and State Vector Updates

The measurements used to track the target are either

$$\mathbf{z}^A = \begin{cases} \|\mathbf{r}\| & \text{Range only Measurement} \\ \left[m_x^A \;\; m_y^A \;\; \|\mathbf{r}\| \right]^T & \text{Range and Monopulse Measurements} \end{cases}.$$

The measurement vector is indicated to be in the antenna frame, but if only the slant range is used, then it is independent of the frame; specifically, $\mathbf{z}^N = \mathbf{z}^A$ in the first case. In the case of three measurements, the measurement vector used in the EKF is

$$\mathbf{z}^N = C_A^N \mathbf{z}^A.$$

In the following steps we will ignore (1) whether a vector is the truth or an estimate or a measurement and (2) that some of the quantities are valid only at a particular time step, instead focusing on the functional dependence. Letting $\mathbf{r}^N = \left[x^N \;\; y^N \;\; z^N \right]^T$ denote the (in reality computed from the estimated) target position of linearization in the ENU frame, the measurement matrix in the N-frame for the range-only case is

$$H_R^N = \left[\frac{\partial \|\mathbf{r}^N\|}{\partial x} \;\; \frac{\partial \|\mathbf{r}^N\|}{\partial y} \;\; \frac{\partial \|\mathbf{r}^N\|}{\partial z} \right] = \left[\frac{x^N}{\|\mathbf{r}^N\|} \;\; \frac{y^N}{\|\mathbf{r}^N\|} \;\; \frac{z^N}{\|\mathbf{r}^N\|} \right] = \left[u_x^N \;\; u_y^N \;\; u_z^N \right],$$

where the last expression is in terms of the estimated unit vector $\widetilde{\mathbf{u}}^N = \widetilde{\mathbf{r}}^N / \|\widetilde{\mathbf{r}}^N\|$ pointed at the target. This is the only H-matrix component needed for the range-only case. In a similar manner, the monopulse measurement matrix components are

$$H_{m_x}^N = \left[\frac{\partial x^N}{\partial x^N} \;\; \frac{\partial x^N}{\partial y^N} \;\; \frac{\partial x^N}{\partial z^N} \right] = [1 \;\; 0 \;\; 0]$$

$$H_{m_z}^N = \left[\frac{\partial z^N}{\partial x^N} \;\; \frac{\partial z^N}{\partial y^N} \;\; \frac{\partial z^N}{\partial z^N} \right] = [0 \;\; 0 \;\; 1].$$

Taking into account the ordering of N-frame measurements relative to the A-frame, the total positional H-matrix is

$$H_k^N = \begin{bmatrix} H_{m_x}^N \\ H_R^N \\ H_R^N \end{bmatrix}.$$

This H-matrix relates position measurements to positions states. In further developments, we are going to track target position and velocity. But the measurements we propose, range or range and monopulse, are all related to position, not velocity. So, the H-matrix for each velocity state H_V^N is just a 1×3 zero matrix:

$$H_V^N = 0_{1 \times 3}.$$

Another decision is needed. One way to track both position and velocity is to order the states as

$$\begin{bmatrix} p_x & v_x & p_y & v_y & p_z & v_z \end{bmatrix}^T,$$

i.e., interspersing position and velocity. This will simplify the forms of the other EKF matrices.

In this interspersed case, the H-matrix becomes

$$H^N = \begin{bmatrix} H_{m_x}^N(1) & 0 & H_{m_x}^N(2) & 0 & H_{m_x}^N(3) & 0 \\ H_R^N(1) & 0 & H_R^N(2) & 0 & H_R^N(3) & 0 \\ H_{m_y}^N(1) & 0 & H_{m_y}^N(2) & 0 & H_{m_y}^N(3) & 0 \end{bmatrix}.$$

It is the ordering of the H-matrix components that defines the ordering of the state vector components. As we have interspersed the H-matrix position and velocity components, the state vector will be interspersed similarly.

There are several components of the EKF that still need definition. The first is the measurement covariance matrix. We are going to assume, as we do in the phase-comparison monopulse simulation, that the noise in the two difference beams is uncorrelated with each other and with the sum beam noise. That makes the R-matrix diagonal, leaving only the two types of measurement covariance terms to be determined. To go further, which we need to do, gets us into the art of Kalman filtering. If R is defined too tightly, meaning set right at the theoretical measurement errors based on knowledge of the SNR, a phenomenon referred to in the literature as "covariance collapse" can occur whereby the EKF position covariance gets too small that in turn makes the Kalman gain too big that then favors the measurements over the filter history. When this happens, measurement fluctuations due to noise or RCS may cause the filter to lose lock on the target.

Our ad hoc approach to covariance collapse is to pretend that both components of the EKF measurement equation

$$\mathbf{Z} = \mathbf{z} - H\delta\mathbf{x},$$

contribute to the measurement error covariance matrix R. Assuming uncorrelated measurement noise and process noise, the covariance of $\mathbf{Z}$ becomes

$$R' = \langle \mathbf{Z}\mathbf{Z}^T \rangle = \langle (\mathbf{z} - H\,\delta\mathbf{x})(\mathbf{z} - H\,\delta\mathbf{x})^T \rangle$$
$$= \langle \mathbf{z}\mathbf{z}^T \rangle + H\langle \delta\mathbf{x}\,\delta\mathbf{x}^T \rangle H^T$$
$$= R_m + HPH^T,$$

where R_m is the actual measurement error covariance, P is our best estimate of the solution error at the current time, and both H and P in this expression are the positional components only (i.e., they do not include the velocity components).

The actual measurement covariance R_m is defined conservatively (i.e., somewhat larger than one thinks the measurement errors to be). To that end, Skolnik (2001) provides estimates of the range measurement error that we rewrite as

$$\sigma_R = \frac{\Delta R}{\sqrt{1 + \mathrm{SNR}}},$$

where ΔR is the range gate size after pulse compression.[1] We have added the one in the denominator to allow this to work at small values of SNR even though when we require that the signal plus noise voltage be detected the difference between SNR and $1 + \mathrm{SNR}$ is small. We also use this expression to generate range measurement errors rather than actually generating the signal plus noise voltages used in the range measurement process with pulse compression included.

Similarly, we set the noise-only phase-comparison measurement error to be the boresight beamwidth based on the results in Fig. 7.2 (eyeball extrapolating the curves at large SNR back toward 0 dB). Thus, we set

$$\sigma_m = \frac{k_{BW}\sqrt{2}}{\sqrt{1 + \mathrm{SNR}}}.$$

Note that the actual monopulse measurements with their errors are generated explicitly from signal plus noise voltages of the sum and difference beams, as described earlier.

The R-matrix then has the form

$$R_m = \begin{cases} \sigma_R^2 & \text{Range-Only Measurement} \\[2mm] \begin{bmatrix} \sigma_m^2 & 0 & 0 \\ 0 & \sigma_R^2 & 0 \\ 0 & 0 & \sigma_m^2 \end{bmatrix} & \text{Range and Monopulse Measurements} \end{cases}.$$

---

[1] In Table 8.1 (on page 64) the pulsewidth τ is 10 μs, giving an uncompressed range bin size $c\tau/2$ of 1500 m. We then set ΔR to 3 m requiring a pulse compression ratio of 500, well within the capability of pulse-compression techniques.

Next, we define the transition matrix Φ. Although we write this as being time sample dependent in Appendix C, we will make it simpler in this analysis, setting the 1-D position/velocity transition matrix to

$$\Phi_{PV} = \begin{bmatrix} 1 & \Delta T \\ 0 & 1 \end{bmatrix},$$

where ΔT is the filter update period. Thus, the position transforms as $p_2 = p_1 + v_1 \Delta T$ and the velocity as $v_2 = v_1$. The full $[6 \times 6]$ transition matrix is block diagonal with three copies of Φ_{PV} along the diagonal.

The process noise is another EKF component that can take a deep dive into the literature. Fortunately, Blackman and Popoli (1999) summarize the Singer model (Singer 1970) that was developed for target acceleration that is a first-order Markov process. In the limit that the maneuver time constant is much less than the sampling period ΔT, in which case acceleration cannot be tracked, a second-order filter is recommended, and the 1-D P-V process noise matrix becomes

$$Q_{PV} = Q_0 \begin{bmatrix} \Delta T^3/3 & \Delta T^2/2 \\ \Delta T^2/2 & \Delta T \end{bmatrix},$$

where Q_0 is a parameter of the problem that has the units $[(m/s^2)^2/Hz]$ of acceleration power spectral density. Blackman and Popoli (1999, §4,2.1) also reiterate the rule of thumb that estimating acceleration is only useful when velocity measurements are available (e.g., by Doppler shift). One could do an optimization study of Q_0 for the tracking problem that we devise to show the benefit of monopulse measurements. But that is beyond the scope of this treatise, and we will use a value of 30 for this parameter.[2]

To obtain monopulse measurements, we need the true AoA and the pointing angle in the antenna frame. Thus, the true angle of arrival vector is

$$\mathbf{k}_{AoA}^A = k_0 C_N^A \mathbf{u}_T^N,$$

where $\mathbf{u}_T^N$ is the unit vector to the true target position at the update time. Similarly, the pointing angle vector is

$$\widetilde{\mathbf{k}}_{Pnt}^A = k_0 C_N^A \widetilde{\mathbf{u}}_k^N,$$

where the unit vector $\widetilde{\mathbf{u}}_k^N$ is based on the estimated target position.

Once the monopulse measurements are obtained and the measurement vector is transformed to the ENU frame, the EKF can be updated. The state vector

---

[2] This is a parameter that should be optimized in a real-world analysis.

components out of the EKF are the corrections to the position $\delta\mathbf{r}_k$ and velocity $\delta\mathbf{v}_k$ in this case. The estimated position and velocity are updated as follows:

$$\widetilde{\mathbf{v}}_k = \widetilde{\mathbf{v}}_k + \delta\mathbf{v}_k$$
$$\widetilde{\mathbf{r}}_k = \widetilde{\mathbf{r}}_k + \widetilde{\mathbf{v}}_k\Delta T + \delta\mathbf{r}_k.$$

There are a number of ways reported in the literature to handle missed detections. Again, this should be a task of finding a fully optimized EKF for one's particular problem. Two obvious choices are (1) to set the Kalman gain to zero when the target is missed (i.e., its sum beam signal plus noise power is below the detection threshold). This keeps the covariance matrix updated through the miss(es), expanding the window in which to look for another detection on subsequent updates. (2) Keep track of the time since the last update after a series of misses, and on the next detection use that time as ΔT in the transition matrix, process noise matrix, and state vector update expressions. One disadvantage of the second technique is that these quantities must be recomputed each time there is a miss (presumably the expressions for the minimum update period can be used otherwise). In this analysis we use the first technique.

A goal of this chapter is to show that the process of tracking one target is not so simple to implement, as indicated by these developments and by the extensive Kalman filter literature.

9.3   Trajectory and Tracking Performance With and Without Monopulse Measurements

To complete this "simple example," we need a trajectory. Our radar can see a Swerling 0 target with a 90 percent detection probability to a range of about 35 km, so we start a track 35 km due north of the radar that is facing north. The target is at 2 km altitude and is flying straight and level at a speed of 200 km/h. Just before overflying the radar, the target makes a 2-min standard turn to the NW, and heads away from the radar. To keep this as simple as possible, the target is above a flat earth. On the first detection, the target is assumed to be anywhere within the 3-dB (one way) beamwidth, as it might be for a search and track radar. As the EKF estimates can wander quite far from the truth after a sequence of misses, the track is dropped after five consecutive misses. Then the target is magically illuminated on the next track update time to restart the track on a detection.

Three cases are presented: EKF with range and monopulse measurements, EKF with just range measurements, and Weighted Least-Squares (WLS, see Appendix C) with range and monopulse measurements.

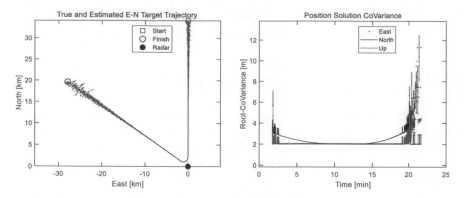

Fig. 9.1 Trajectory and EKF position covariance with range and monopulse measurements

Case 1: EKF With Range and Monopulse Measurements

This is the best tracking performance case of the three. The East-North trajectory is plotted in the left frame of Fig. 9.1. The trajectory start and end points are indicated, and the radar is at the origin of the ENU coordinate system. Below the dots there is a black line that indicates the true trajectory, but this is overlaid by blue and red dots (for detections and misses, respectively) indicating the estimated position of the EKF. The misses are much more prominent at the end of the track than at the beginning for reasons to be explained.

The position root-covariance of the EKF solution is plotted in the right frame of Fig. 9.1. As the East (blue dots) and Up (magenta lines) directions correspond to monopulse measurements, these fall on top of one another. The back line is the North covariance that corresponds to the North direction for a target on boresight (which it never is exactly true in this analysis).

The spikes in the covariance at either end of the trajectory are the result of missed detections. While in the region where the detection probability is essentially unity, between about 2.7 and 19 min, the East and Up root-covariance is 2.04 m. In a smaller time interval, the North root-covariance has the same value.

If we use the true R-matrix (scaled by 1/SNR without the additional term to prevent covariance collapse) or a constant R-matrix corresponding to unity *SNR*, the minimum root-covariance is 0.4 m that causes continuous missed detection (or loss of lock) events. Mitigating such behavior is part of the art of Kalman filtering, a subject beyond the scope of this work.

In Fig. 9.2 are plots of signal-to-noise ratios (left frame) and defining angles (right frame). The "true SNR" is that of the signal *if* the AESA beam were centered on the target. The actual SNR (blue or red dots for detections or not, respectively) differs from the truth because the AESA face is vertical, and as the target approaches the radar it appears higher in the sky and its return suffers cosine taper loss entering the elements. Then as the target turns away, its bearing is 55-degrees from North, resulting in similar taper loss. The trajectories of the AoA angle from boresight are

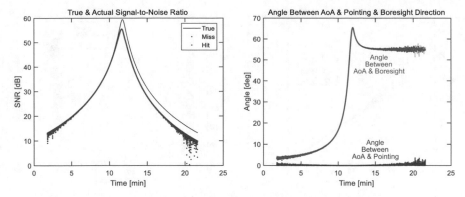

Fig. 9.2 Signal-to-Noise Ratio and defining angles versus time along trajectory

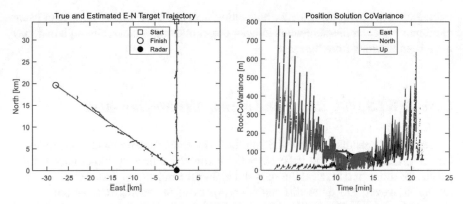

Fig. 9.3 Trajectory and EKF position covariance with only range measurements

plotted with magenta (hits) and cyan (misses) dots, that overlay the true trajectory of this angle. The largest angle from boresight occurs at the closest range when the target is high in the sky and turning to the new heading. At this point, the angle from boresight is about 65-degrees, and the taper loss is about 5 dB.

At the bottom of the right plot is the angle between the true AoA and the pointing angle based on the estimated position. For most of the trajectory, this pointing error is much smaller than the 3-dB beamwidth of 3°.

Case 2: EKF With Only Range Measurements

The East-North trajectory and estimated position are plotted in the left frame of Fig. 9.3. It is clear that the EKF struggles to maintain lock on the target with missed detections and reacquisition events in numerous spots along the trajectory where the 3-measurement EKF had none. The root-covariance plotted in the right frame shows

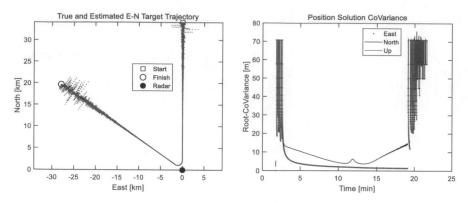

Fig. 9.4 Trajectory and WLS position covariance with range and monopulse measurements

track-lose lock-reacquire sequence happening over and over. The advantage of using monopulse angular measurements in radar target tracking is clear, although that may have been obvious from the start.[3]

Case 3: WLS With Range and Monopulse Measurements

Corresponding results for the WLS solution with three measurements are plotted in Fig. 9.4. Compared to the EKF case, the estimated trajectory looks very similar except for more "red fuzz" at the start and end of the trajectory. The root-covariance results, however, are quite different (larger), even in regions not suffering from missed detections. The cost of the simpler WLS solution is larger errors and covariance terms. So, in most situations, particularly with modern general-purpose processors, the full EKF should be implemented.

---

[3] An extended Kalman filter presents its builder with any number of opportunities to get things wrong. Sometimes the filter can correct relatively minor infractions, but sign errors, for example, almost always end badly. So, one must be careful that such bad tracking is the actual situation and not just a reflection of an error in the EKF formulation. It was not until the three-measurement results were obtained (which took a number of code iterations) that these results were accepted as what really can happen.

Chapter 10
Fidelity and Accuracy of Monopulse Measurement Simulation

Abstract The process of simulating monopulse measurements with an AESA brought about a number of decisions, some which were later changed, as to what should be included and what should not. This is a process that all simulation developers undergo, sometimes tacitly, when starting work. As one cannot simulate the universe and meet deadlines and budgets, spending time on this up front often will produce a better product with less "wasted effort," although this is in quotes because often discovering what is wrong or not well understood or unimportant is indeed "productive effort" in simulation development.

Keywords Monte Carlo simulation · Simulation fidelity · Simulation accuracy

> *"All models are wrong, but some are useful." — George F. P. Box.*

10.1 Fidelity of Monopulse Simulation Model

The process of designing an AESA starts with the structure of an element. The tools used for this design effort are first principles electromagnetic solvers such as High Frequency Structure Simulator (HFSS). Even then, the element often is assumed to be in an array of elements extending to infinity in the array plane to simplify the boundary conditions. Despite this extent, it is not possible to simulate the far-field gain of an array because of the problem of scale—resolution much less than a wavelength to do justice to the element structure, but with a much larger extent in the far field. These two are not compatible.

Our AESA simulation starts with modeling an element as radiating or receiving plane waves, with the gain of the element versus direction from boresight modeled with the cosine taper function. Its efficiency at these tasks is assumed to be perfect, so we make no distinction between directivity and gain.

Complicating effects such as mutual coupling between elements or scan blindness, for example, are ignored, as a well-designed AESA will minimize these effects within its required scan volume. Impedance mismatch with scan angle is accounted

© The Author(s), under exclusive license to Springer Nature Switzerland AG 2022
R. A. Dana, *Monopulse Measurement with Active Electronically Scanned Arrays (AESAs)*, https://doi.org/10.1007/978-3-030-91908-5_10

for in the cosine taper exponent being larger than unity, and again this is minimized by good element design. Other effects that we do include are element failures, quantization, and vector modulator noise. These are real issues in AESA design and are readily modeled within the context of our simulation by their effect on element weights.

So what simulation fidelity is sufficient to make our model useful? The answer depends on the specific questions, of course. The questions of this work are about the operation of monopulse using an AESA, and important issues are those that affect the phase and amplitude of the received signals. However, if this were work being done for a project, then maybe the most important question is "Does the design meet the requirements?" Here is where things can get tricky. Under what conditions? Pristine new system, end of life, been around the block a few times, measured in the lab, simulated? Many of these do not lend themselves to pure analysis or simulation. What we have presented is "ideal performance" even in the presence of element failures, quantization, and VM noise.

Chapter 8 has examples of how to include atmospheric attenuation in computing a detection range. This was not done in the AESA simulation; rather it was computed with a side code that is deterministic, not Monte Carlo, implementing the rain and air attenuation algorithms for the standard atmosphere that are available from NOAA and ITU. The AESA was assumed to be ideal, and the target was placed at the peak of the beam.

10.2   Simulation Accuracy

There are at least two levels of accuracy that should be addressed in developing a simulation and presenting its results. The first is: does it give the right answer? Of course, George Box deflected that question to one of usefulness. But we have shown three examples of where the simulation matches the truth as defined by analytic expressions. The first was the approximation reported in Sherman and Barton (2011) that the on-axis phase-comparison StD error goes as $\Theta_0/\sqrt{2SNR}$. The second was the agreement of the amplitude-comparison monopulse error as a function of squint angle and of SNR at the optimum squint angle shown in Fig. 5.4. Here, the agreement was exact until the analytic expressions became inaccurate because of their dependence on a Gaussian-shaped beam. The third comparison was between the simulated probability of detection and the Swerling detection formulas shown in Fig. 7.1.

In addition to these obvious triumphs (none of which came out correctly on first try), there are the countless number of results, either printed or plotted during simulation development that one must examine: Does this make sense? Does it agree with my intuition? Do I understand what I am seeing? If, not, stop and take the time to understand what is happening. It may be intuition is incorrect, but more often than not it is something wrong in the simulation.

Once one has confidence in the simulation results, the other level of accuracy is in the number of Monte Carlo samples. The reader may have noticed that this has not

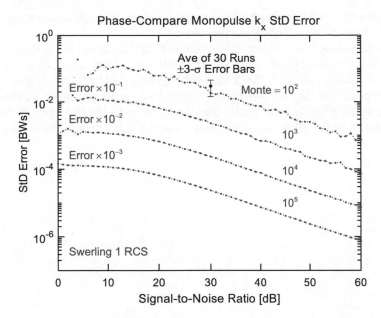

Fig. 10.1 Radar phase-comparison monopulse error versus SNR for various numbers of Monte Carlo Samples

been mentioned before, because the author is a stickler for smooth (i.e., not noticeably jagged) simulation curves. Generally, the number of Monte Carlo samples is increased until either the curves appear smooth or the CPU time to run the cases gets too long. Most of the simulation results, which includes an entire curve, were run in a few minutes to about 20 min (e.g., the VM contour plots). This is achieved, in part, by limiting the calculations within the Monte Carlo loop to only those that depend on Monte Carlo variables.

The Swerling 1 radar phase-comparison monopulse measurement error is a good example of the dependence of the error on the number of Monte Carlo samples, referred to as "Monte," in Fig. 10.1. The top curve is for Monte of 100, and the lower curves are for Monte of 10^3, 10^4, and 10^5, respectively. Each of these curves has been shifted down by a factor of 10 so the curves do not fall on top of one another. The curve with 100,000 Monte Carlo samples and 61 SNR values took only about 2 min to compute on a Dell Precision laptop, well within the author's "waiting for it to complete, tapping his fingers" tolerance. These curves illustrate nicely the "not noticeably jagged" criterion that only the bottom curve fulfills across the range of the abscissa.

The top curve at 30 dB has a 3σ error bar about an average (black point) computed for 30 different Monte Carlo runs, each with 100 Monte Carlo samples. The error bars encompass the other (blue) point at 30 dB, but not by much. So, what if your problem or simulation code takes much longer than a couple of minutes to generate the 100 Monte Carlo sample curve and you cannot afford to generate the 1000 sample, 61-SNR value curve? Running a number of cases at the lower number of Monte Carlo samples allows one to compute the mean and standard deviation of the results at one SNR value. Now one has an idea of the Monto Carlo accuracy of the original curve with a limited number of samples, something one does not get by just upping Monte and generating another curve.

Where having this error bar can become critically important is when the curve is close to the required performance at some SNR value, for example. Now one could generate a histogram of say 100 results and hopefully declare that the requirement is met 99 percent of the time, something program managers and customers like to hear. However, keep in mind that in a lab the signal power of a test setup can be uncertain too, maybe by 0.5 dB or so, there is also a horizontal error bar in the real world.

Appendix A: Amplitude-Comparison Monopulse Squinting in Two Dimensions

Determining the best squint angle directions is less trivial than it may seem at first glance. For example, if the beam is pointed at boresight, then the $x-$ and $y-$directions would seem to be best as this is the most natural coordinate systems of the AESA. But say we are trying to point in the $x-$direction at the maximum allowed elevation angle. Then the best directions of measurement might be in the $x-$ and $y-$directions, or they might be in the azimuth/elevation angle directions. But if we again want to point at boresight, the azimuthal direction is indeterminate, and it turns out that the required azimuth angle offsets are too large near this pointing direction. Of course, at boresight rotating two beams in different azimuthal directions does not achieve any angular separation. So, we need to approach this problem without a single set of direction coordinates in mind, and we consider both rectangular and polar coordinates.

There is another limitation, one that is inherent in electronically scanned arrays, to consider. We should not scan the elevation angle to 90° because the signal we want to receive at this angle is propagating parallel to the array, and the effective area of the elements as seen by the wave is zero, resulting in a theoretical array gain of zero. A design-specific scan limit is say 60°, an angle where the impedance mismatch between the elements and the free space signal is not too large.

Squinting Increments in Cartesian Coordinates

A fairly tedious exercise in trigonometry, made possible by Mathematica, is required to determine the amplitude-comparison monopulse squint angles in two dimensions. We point the sum beam as indicated in Fig. 3.1, with elevation angle θ_0 and azimuth angle φ_0:

$$\mathbf{k}_0 = k_0[\sin\theta_0\cos\varphi_0\widehat{\mathbf{x}} + \sin\theta_0\sin\varphi_0\widehat{\mathbf{y}} + \cos\theta_0\widehat{\mathbf{z}}],$$

where $k_0 = 2\pi/\lambda$ is the magnitude of $\mathbf{k}_0$, and the unit vectors $\widehat{\mathbf{x}}$ and $\widehat{\mathbf{y}}$ are parallel to the two sides of the AESA in such a way that the unit vector $\widehat{\mathbf{z}}$ is normal to the face of the array and points along its boresight.

Then to do amplitude-comparison monopulse in two orthogonal $x-$ and $y-$directions, four squinted beams must be formed that we denote by their pointing directions $\mathbf{k}_{Sx,\;+}$, $\mathbf{k}_{Sx,\;-}$, $\mathbf{k}_{Sy,\;+}$, and $\mathbf{k}_{Sy,\;-}$, where the $+/-$ beams are pointed away from $\mathbf{k}_0$ in the positive/negative directions:

$$\mathbf{k}_{Sx,+} = k_0[(\sin\theta_0\cos\varphi_0 + \Delta_{Sx+})\widehat{\mathbf{x}} + \sin\theta_0\sin\varphi_0\widehat{\mathbf{y}} + k_{z,Sx+}\widehat{\mathbf{z}}]$$
$$\mathbf{k}_{Sx,-} = k_0[(\sin\theta_0\cos\varphi_0 - \Delta_{Sx-})\widehat{\mathbf{x}} + \sin\theta_0\sin\varphi_0\widehat{\mathbf{y}} + k_{z,Sx-}\widehat{\mathbf{z}}]$$
$$\mathbf{k}_{Sy,+} = k_0[\sin\theta_0\cos\varphi_0\widehat{\mathbf{x}} + (\sin\theta_0\sin\varphi_0 + \Delta_{Sy+})\widehat{\mathbf{y}} + k_{z,Sy+}\widehat{\mathbf{z}}]$$
$$\mathbf{k}_{Sy,-} = k_0[\sin\theta_0\cos\varphi_0\widehat{\mathbf{x}} + (\sin\theta_0\sin\varphi_0 - \Delta_{Sy-})\widehat{\mathbf{y}} + k_{z,Sy-}\widehat{\mathbf{z}}].$$

The z components are variables to be determined that keep the magnitudes of these direction vectors equal to k_0. The other constraint is that the angular difference between these vectors and $\mathbf{k}_0$ must be equal to $\theta_S/2$. This is imposed by noting that the dot product of unit vectors is equal to the cosine of the angle between them. Thus, for example,

$$\frac{\mathbf{k}_{Sx,+} \cdot \mathbf{k}_0}{k_0^2} = \cos(\theta_S/2).$$

Similar expressions hold for the other three squint direction vectors.

Setting the magnitudes of the squint pointing vectors to unity gives expressions for the $z-$components in terms of the Δ unknowns:

$$k_{z,Sx+} = \sqrt{\cos^2\theta_0 - 2\sin\theta_0\cos\varphi_0\Delta_{Sx+} - \Delta_{Sx+}^2}$$

$$k_{z,Sx_-} = \sqrt{\cos^2\theta_0 + 2\sin\theta_0\cos\varphi_0\Delta_{Sx-} - \Delta_{Sx-}^2}$$

$$k_{z,Sy+} = \sqrt{\cos^2\theta_0 - 2\sin\theta_0\sin\varphi_0\Delta_{Sy+} - \Delta_{Sy+}^2}$$

$$k_{z,Sy-} = \sqrt{\cos^2\theta_0 + 2\sin\theta_0\sin\varphi_0\Delta_{Sy-} - \Delta_{Sy}^2}\;.$$

Then the angular distance requirement results in quadratic equations, the solutions of which are

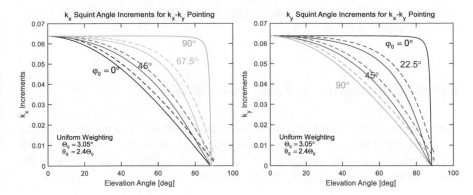

Fig. A.1 k_x and k_y squint angle increments versus elevation angle

$$\Delta_{Sx+} = \frac{\sqrt{2}\cos\theta_0\sin(\frac{\theta_S}{4})\sqrt{\cos^2\theta_0 + \cos(\frac{\theta_S}{2}) + \cos(2\varphi_0)\sin^2\theta_0 - 2\cos\varphi_0\sin\theta_0\sin^2(\frac{\theta_S}{4})}}{\cos^2\theta_0 + \cos^2\varphi_0\sin^2\theta_0}$$

$$\Delta_{Sx-} = \frac{\sqrt{2}\cos\theta_0\sin(\frac{\theta_S}{4})\sqrt{\cos^2\theta_0 + \cos(\frac{\theta_S}{2}) + \cos(2\varphi_0)\sin^2\theta_0 + 2\cos\varphi_0\sin\theta_0\sin^2(\frac{\theta_S}{4})}}{\cos^2\theta_0 + \cos^2\varphi_0\sin^2\theta_0}$$

$$\Delta_{Sy+} = \frac{\sqrt{2}\cos\theta_0\sin(\frac{\theta_S}{4})\sqrt{\cos^2\theta_0 + \cos(\frac{\theta_S}{2}) - \cos(2\varphi_0)\sin^2\theta_0 - 2\sin\varphi_0\sin\theta_0\sin^2(\frac{\theta_S}{4})}}{\cos^2\theta_0 + \sin^2\varphi_0\sin^2\theta_0}$$

$$\Delta_{Sy-} = \frac{\sqrt{2}\cos\theta_0\sin(\frac{\theta_S}{4})\sqrt{\cos^2\theta_0 + \cos(\frac{\theta_S}{2}) - \cos(2\varphi_0)\sin^2\theta_0 + 2\sin\varphi_0\sin\theta_0\sin^2(\frac{\theta_S}{4})}}{\cos^2\theta_0 + \sin^2\varphi_0\sin^2\theta_0}$$

In the limit of boresight pointing ($\theta_0 = \varphi_0 = 0$), all four expressions reduce to:

$$\Delta_{Sx+} = \Delta_{Sx-} = \Delta_{Sy+} = \Delta_{Sy-} = \sin(\theta_S/2).$$

Figure A.1 has plots of the k_x (left frame) and k_y (right frame) squint angle increments (solid lines are in the positive direction; dashed lines in the negative direction), normalized to $\sin(\theta_S/2)$, for various pointing azimuth angles $\varphi_0 = 0°$, $22.5°$, $45°$, $67.5°$, and $90°$. (One of these is omitted in each frame for clarity.) The plot abscissa ranges[1] from $0°$ to $90°$, whereas a real AESA would not scan in elevation

---

[1] The extension of the plots to $90°$ elevation angle is pedagogical. We will see that squinting in the azimuth and elevation angle directions is preferred for elevation angles above $3°$ or so.

beyond $60°$ or so. These results are peculiar to our 32×32 element array with an unweighted beamwidth of $3.05°$.

Squinting Increments in Polar Coordinates

Calculating the squint angle increments in polar coordinates proceeds in much the same manner as above, except that the increments are in the elevation and azimuth angles rather than in the rectangular components of $\mathbf{k}$. The elevation angle increments are obtained by setting the dot products of

$$\mathbf{k}_{S\theta} = k_0[\sin(\theta_0 + \Delta_\theta)\cos\varphi_0\mathbf{x} + \sin(\theta_0 + \Delta_\theta)\sin\varphi_0\mathbf{y} + \cos(\theta_0 + \Delta_\theta)\mathbf{z}]$$
$$\mathbf{k}_{S\varphi} = k_0[\sin(\theta_0)\cos(\varphi_0 + \Delta_\varphi)\mathbf{x} + \sin(\theta_0)\sin(\varphi_0 + \Delta_\varphi)\mathbf{y} + \cos(\theta_0)\mathbf{z}]$$

and $\mathbf{k}_0$ to

$$\frac{\mathbf{k}_0 \cdot \mathbf{k}_{S\theta}}{k_0^2} = \cos(\theta_S/2) \quad \frac{\mathbf{k}_0 \cdot \mathbf{k}_{S\varphi}}{k_0^2} = \cos(\theta_S/2),$$

which gives

$$\Delta_\theta = \pm\theta_S/2$$
$$\Delta_\varphi = \pm\cos^{-1}\left[\left(\cos(\theta_S/2) - \cos^2\theta_0\right)\csc^2\theta_0\right].$$

These polar coordinate increments are plotted in the left frame of Fig. A.2 versus elevation pointing angle. The dashed line at $\theta_0 = 1.84°$ indicates where the azimuth increment reaches $180°$, at which point it becomes ambiguous. In the right frame of the figure are the errors that are made in the pointing directions of the squint beams

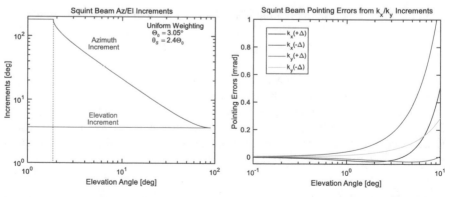

Fig. A.2 Polar coordinate squint angle increments (left frame) and rectangular increment errors versus elevation angle

using increments computed in the rectangular $x-$ and $y-$directions.[2] There errors are small (less than 0.2 mrad) for elevation angles less than 3 degrees, which gives a threshold for the two types of increments: rectangular increments for an elevation angle less than 3 degrees (or one beamwidth), and polar increments for larger elevation angles. Of course, the increments for both polar and rectangular coordinates depend on the beamwidth and on our choice of the squint angle.

---

[2]These errors occur because we compute the rectangular coordinate increments in one dimension assuming that the other-dimension increment is zero, whereas we would like the increments to move the squint beams along one arc from the main beam.

Appendix B: A Property of Uniformly Weighted AESA Gain in One Dimension

Dana (2019) gives the 1-D gain for an AESA with N_e elements and element spacing of $a\lambda$ ($a = 1/2$ in our case with $\lambda/2$ element spacing) as a function of un-normalized 1-D k-vector magnitude

$$G(k) = (a\lambda)^2 \frac{\sin^2[a\lambda N_e k/2]}{N_e(a\lambda k/2)^2}.$$

The first null in the pattern occurs where the sine argument is equal to π, or where the 1-D normalized $\widetilde{k}$ is equal to

$$\widetilde{k}_{Null} = \frac{k_{Null}}{2\pi/\lambda} = \frac{2}{N_e}.$$

The Full-Width, Half Maximum (FWHM) beamwidth of the gain is where

$$G(k_{BW}/2) = \frac{G(0)}{2}.$$

The first positive-argument solution to the equation $\sin^2(x_0) = x_0^2/2$ is $x_0 = 1.39156\cdots$, which gives the normalized k-space beamwidth:

$$\widetilde{k}_{BW} = \frac{4x_0}{\pi N_e}.$$

Thus, the first positive-k null, measured in 3-dB beamwidths occurs at

$$\frac{\widetilde{k}_{Null}}{\widetilde{k}_{BW}} = \frac{\pi}{2x_0} = 1.1288\cdots$$

© The Author(s), under exclusive license to Springer Nature Switzerland AG 2022
R. A. Dana, *Monopulse Measurement with Active Electronically Scanned Arrays (AESAs)*, https://doi.org/10.1007/978-3-030-91908-5

Appendix C: "Derivation" of the Linear Kalman Filter, Extended Kalman Filter, and Weighted Least-Squares Equations

One way to understand a discrete Kalman Filter (i.e., updated at regular time intervals $k\,\Delta T$) and its limitations is to "derive" its equations from the basic principles from which it is formulated. Derive is in quotes because this is not a formal mathematical derivation; rather it is more of a listing of the assumptions underlying the filter and their consequences. There are three parts to these developments. First is the Linear Kalman Filter (LKF) wherein there is a linear relationship between the states that are tracked and the measurements that are used to update the filter. Second is the Extended Kalman Filter (EKF) wherein there is a mathematical relationship between the states and the measurements that one can make into a linear relationship by expanding about a "linearization point" in a first-order Taylor series. Third is the Weighted Least-Squares (WLS) solution that is a subset of the EKF that does not carry as much "history" along.

Linear Kalman Filter

The basis of the Kalman filtering is two "first principles" equations that describe the evolution of the process state vector $\mathbf{x}_k$ (subsequently referred to as the state vector) that represents "the stuff we want to keep track of," and the linear relationship between the states and the measurements $\mathbf{z}_k$:

$$\boxed{\begin{aligned} \mathbf{x}_{k+1} &= \Phi_k \mathbf{x}_k + \mathbf{w}_k \\ \mathbf{z}_k &= H_k \mathbf{x}_k + \mathbf{v}_k \end{aligned}}\ \text{(LKF)}, \qquad (C.1)$$

R. A. Dana, *Monopulse Measurement with Active Electronically Scanned Arrays (AESAs)*, https://doi.org/10.1007/978-3-030-91908-5

where here the subscript k indicates the quantities at time $t_k = k\Delta T$, and $\mathbf{x}_k = [n \times 1]^3$ process state vector at time t_k, $\mathbf{\Phi}_k = [n \times n]$ transition matrix,[4] $\mathbf{w}_k = [n \times 1]$ vector of white noise samples, $\mathbf{z}_k = [m \times 1]$ vector of measurements made at time t_k, $H_k = [m \times n]$ matrix relating measurements to the state vector, $\mathbf{v}_k = [m \times 1]$ vector of measurement errors at time t_k.

The number of measurements m may be less than, equal to, or greater than the number of states n. A rule of thumb of Kalman filtering is that the order of the state vector should be no larger than the order of the measurements plus one. For example, in radar if we measure slant range to the target and two monopulse angles, all measurements are first order (i.e., related to position). Then the Kalman filter should track position and velocity but not acceleration. The reason is that acceleration in this case is obtained by double differencing position, and each difference adds more noise to the acceleration estimate.

The covariance matrices of the noise processes $\mathbf{w}_k$ and $\mathbf{v}_k$ are given by

$$\left\langle \mathbf{w}_k \mathbf{w}_l^T \right\rangle = Q_k \delta_{k,l}$$
$$\left\langle \mathbf{v}_k \mathbf{v}_l^T \right\rangle = R_k \delta_{k,l}$$
$$\left\langle \mathbf{w}_k \mathbf{v}_l^T \right\rangle = 0,$$

where the superscript T denotes the matrix transpose, and $\delta_{k,\,l}$ is the Kronecker delta symbol. The matrices Q and R will be defined later when we apply the general results derived here to a specific radar problem.

Note that the state vector noise is an artifact of the filter principles; it does not mean that there is actual noise that somehow appears in each state vector update. Rather, its purpose is to indicate that there is some uncertainty in the update dynamics defined later by the deterministic transition matrix. For example, an airborne target flying straight and level for some time might suddenly turn. The transition matrix may not account for acceleration or jerk (the rate of change of acceleration), and the state vector noise allows for the extra dynamics in the target between updates that are not described by the transition matrix. On the other hand, measurements are always accompanied by noise that we assume is zero-mean, normally distributed.

Assume that we have an initial estimate of the process at time t_k based on our knowledge about the process for $t < t_k$. This estimate is denoted by $\widetilde{\mathbf{x}}_k^-$, where the "tilde" denotes an estimate of the vector $\mathbf{x}_k$, and the superscript "$-$" denotes our estimate before the assimilation of the measurements taken at time t_k. Also assume that we know the covariance associated with the error in $\widetilde{\mathbf{x}}_k^-$. Denoting this error as $\mathbf{e}_k = \mathbf{x}_k^- - \widetilde{\mathbf{x}}_k^-$, the truth minus our estimate of the truth before incorporating the current measurements, the $[n \times n]$ error covariance matrix is

---

[3] The notation indicates the number of rows (n) by the number of columns (1 in this case).

[4] Upper case characters denote matrices.

$$P_k^- = \langle \mathbf{e}_k^- \mathbf{e}_k^{-T} \rangle = \left\langle \left(\mathbf{x}_k^- - \widetilde{\mathbf{x}}_k^- \right) \left(\mathbf{x}_k^- - \widetilde{\mathbf{x}}_k^- \right)^T \right\rangle,$$

where we tacitly assume that

$$\langle \mathbf{e}_k^- \rangle = 0.$$

Now we seek to use the current measurements $\mathbf{z}_k$ to improve our estimate of the current state:

$$\widetilde{\mathbf{x}}_k = \widetilde{\mathbf{x}}_k^- + K_k(\mathbf{z}_k - H_k\widetilde{\mathbf{x}}_k^-)(\text{LKFEq.2}), \qquad (\text{C.2})$$

where $\widetilde{\mathbf{x}}_k$ is the updated (a posteriori) estimate of the state vector, the $[n \times m]$ matrix K_k is the Kalman gain to be computed, and the term in parentheses represents the difference between the actual measurements and what we would predict them to be based on our most recent state vector. This difference, modified by the KF gain, is used to update the state vector. This is the second equation to be evaluated in the steps of updating the filter at each time step.

The Kalman gain is determined by minimizing the mean-square error in $\widetilde{\mathbf{x}}_k$:

$$P_k = \langle \mathbf{e}_k \mathbf{e}_k^T \rangle = \left\langle (\mathbf{x}_k - \widetilde{\mathbf{x}}_k)(\mathbf{x}_k - \widetilde{\mathbf{x}}_k)^T \right\rangle.$$

Using the expression in Eq. (C.2) for $\widetilde{\mathbf{x}}_k$ and expanding the result, we get eight terms:

$$P_k = \underbrace{\left\langle (\mathbf{x}_k - \widetilde{\mathbf{x}}_k^-)(\mathbf{x}_k - \widetilde{\mathbf{x}}_k^-)^T \right\rangle}_{1} - \underbrace{\left\langle (\mathbf{x}_k - \widetilde{\mathbf{x}}_k^-)(\mathbf{x}_k - \widetilde{\mathbf{x}}_k^-)^T \right\rangle H_k^T K_k^T}_{2}$$

$$- \underbrace{\left\langle (\mathbf{x}_k - \widetilde{\mathbf{x}}_k^-)\mathbf{v}_k^T \right\rangle K_k^T}_{3} - \underbrace{K_k H_k \left\langle (\mathbf{x}_k - \widetilde{\mathbf{x}}_k^-)(\mathbf{x}_k - \widetilde{\mathbf{x}}_k^-)^T \right\rangle}_{4}$$

$$+ \underbrace{K_k \left\langle \mathbf{v}_k (\mathbf{x}_k - \widetilde{\mathbf{x}}_k^-)^T \right\rangle}_{5} + \underbrace{K_k H_k \left\langle (\mathbf{x}_k - \widetilde{\mathbf{x}}_k^-)(\mathbf{x}_k - \widetilde{\mathbf{x}}_k^-)^T \right\rangle H_k^T K_k^T}_{6}$$

$$+ \underbrace{K_k \left\langle \mathbf{v}_k (\mathbf{x}_k - \widetilde{\mathbf{x}}_k^-)^T \right\rangle H_k^T K_k^T}_{7} + \underbrace{K_k \left\langle \mathbf{v}_k \mathbf{v}_k^T \right\rangle K_k^T}_{8}$$

The first term on the right-hand side of this equation is just P_k^-, the covariance extrapolated to the kth time step from the $(k-1)^{th}$ step. Noting from our definitions of the error, $\mathbf{e}_k$ must be uncorrelated with the current zero-mean measurement noise $\mathbf{v}_k$ because the truth $\mathbf{x}_k$ does not depend on the measurements, and $\widetilde{\mathbf{x}}_k^-$ is the estimated state before incorporation of the current measurement. So, terms 3, 5, and 7 are identically zero. Then, this expression becomes

$$P_k = (I_n - K_k H_k) P_k^-(I_n - K_k H_k)^T + K_k R_k K_k^T, \tag{C.3}$$

where I_n is the $[n \times n]$ identity matrix. The optimization problem is to find the matrix K_k that minimizes the diagonal terms of P_k. To this end we expand P_k into a polynomial in K_k,

$$P_k = P_k^- - K_k H_k P_k^- - P_k^- H_k^T K_k^T + K_k \left(H_k P_k^- H_k^T + R_k\right) K_k^T$$

and use the following matrix identities:

$$\frac{d}{dA}[\text{trace}\,(AB)] = B^T$$

$$\frac{d}{dA}\left[\text{trace}\,(A\,CA^T)\right] = 2AC.$$

where AB is a square matrix and C is a symmetric matrix. Also, the derivative of a scalar s with respect to a matrix is defined as

$$\frac{ds}{dA} = \begin{bmatrix} ds/da_{1,2} & ds/da_{1,2} & \cdots \\ ds/da_{2,1} & ds/da_{2,2} & \\ \vdots & & \ddots \end{bmatrix}.$$

We want to minimize the trace of P_k because this is the sum of the mean-square errors in the estimate of the elements of the state vector. We argue that the individual mean-square errors are minimized when the total is minimized provided that we have sufficient degrees of freedom in the gain matrix. Taking the derivative of the trace of P_k with respect to K_k and setting the result to zero,

$$\frac{d(\text{trace}\,P_k)}{dK_k} = -2\left(H_k P_k^-\right)^T + 2K_k\left(H_k P_k^- H_k^T + R_k\right) = 0,$$

gives the following for the Kalman gain matrix:

$$K_k = P_k^- H_k^T \left(H_k P_k^- H_k^T + R_k\right)^{-1} \text{ (LKF Eq.1).} \tag{C.4}$$

Inserting this expression into Eq. (C.3) and expanding the result gives the following equation that is used to update P_k from P_k^-:

$$P_k = (I - K_k H_k) P_k^- \text{ (LKF Eq.3).} \tag{C.5}$$

This form does not necessarily preserve the inherent symmetry of the covariance matrix, and asymmetry can build up over a large number of updates from just numerical noise. So, often Eq. (C.5) is followed by the operation

Table C.1 Linear Kalman filter update equations

LKF equation number	LKF equation (In order of update)
1	$K_k = P_k^- H_k^T \left(H_k P_k^- H_k^T + R_k \right)^{-1}$
2	$\widetilde{\mathbf{x}}_k = \widetilde{\mathbf{x}}_k^- + K_k \left(\mathbf{z}_k - H_k \widetilde{\mathbf{x}}_k^- \right)$
3a	$P_k = (I - K_k H_k) P_k^-$
3b	$P_k = \left(P_k + P_k^T \right)/2$
4	$\widetilde{\mathbf{x}}_{k+1}^- = \Phi_k \widetilde{\mathbf{x}}_k$
5	$P_{k+1}^- = \Phi_k P_k \Phi_k^T + Q_k$

$$P_k = \left(P_k + P_k^T \right)/2$$

to force symmetry.

Now we are able to project our estimated state vector ahead to the next time t_{k+1} using the deterministic transition matrix:

$$\widetilde{\mathbf{x}}_{k+1}^- = \Phi_k \widetilde{\mathbf{x}}_k \text{ (LKF Eq.4).} \tag{C.6}$$

The error in the covariance matrix associated with $\widetilde{\mathbf{x}}_{k+1}^-$ is

$$\mathbf{e}_{k+1}^- = \mathbf{x}_{k+1} - \widetilde{\mathbf{x}}_{k+1}^- = \Phi_k \mathbf{e}_k + \mathbf{w}_k.$$

Note that $\left\langle \mathbf{e}_k \mathbf{w}_k^T \right\rangle = 0$ because $\mathbf{w}_k$ is the process noise for updating the state vector to time step t_{k+1}. Thus, the projected error covariance matrix is

$$P_{k+1}^- = \left\langle \mathbf{e}_{k+1}^- \mathbf{e}_{k+1}^{-T} \right\rangle = \Phi_k \left\langle \mathbf{e}_k \mathbf{e}_k^T \right\rangle \Phi_k^T + Q_k.$$

The second expectation is just P_k, so this equation reduces to

$$P_{k+1}^- = \Phi_k P_k \Phi_k^T + Q_k \text{ (LKF Eq.5).} \tag{C.7}$$

Equations (C.2), (C.4), (C.5), (C.6), and (C.7), in order, comprise the LKF recursive equations that are listed again in Table C.1.

Extended Kalman Filter

The problem with the LKF for almost all practical applications is the assumption of the linear relationship between the Kalman filter states and the measurements. For example, in a radar the measurements are the slant range to the target and the monopulse measurements of the target displacement from the pointing direction. The states might be the target position in an East-North-Up (ENU) coordinate system with the radar at the origin. This is the problem we consider in Chap. 9.

For the EKF, the underlying equations are

$$\begin{matrix} \mathbf{x}_k = \Phi_{k-1}\mathbf{x}_{k-1} + \mathbf{w}_{k-1} \\ \mathbf{z}_k = \mathbf{h}(\mathbf{x}_k) + \mathbf{v}_k \end{matrix} \quad \text{(EKF)}, \qquad \text{(C.8)}$$

where $\mathbf{h}(\mathbf{x}_k)$ is the known and differentiable functional relationships between the states and the measurements. Note that the time indices in the first equation have been modified from the LKF for reasons explained below.

As an example of the function $\mathbf{h}(\mathbf{x}_k)$, the slant range of a radar target is the vector norm of the positional part of the state vector, denoted by ξ_k, $^5 R = \|\xi_k\|$. Also, in the radar we consider, the measurements include the monopulse angular displacements of the target from the array pointing direction, again related in a nonlinear way to the estimated target position.

The way to linearize the measurement equation, which is necessary for the other Kalman filter equations to hold, is to expand the equation about a "linearization point" $\mathbf{x}_0 = \widetilde{\mathbf{x}}_{k-1}$ that is understood to be the best estimate of the state vector at the last update (time sample $k - 1$). Then the measurement equation becomes

$$\mathbf{z}_k = \mathbf{h}(\mathbf{x}_0) + \left(\frac{\partial \mathbf{h}}{\partial \mathbf{x}}\right)_{\mathbf{x}=\mathbf{x}_0} \cdot \delta\mathbf{x}_k + HOTs.$$

Here *HOTs* are higher-order terms that are ignored, and $\delta\mathbf{x}_k$ is the error in the linearization point that is the quantity that the EKF will produce. The linearized measurement equation is a rearrangement of this equation, and the "first-principles" EKF equations become

$$\boxed{\begin{matrix} \delta\mathbf{x}_k = \Phi_k\delta\mathbf{x}_{k-1} + \mathbf{w}_{k-1} \\ \mathbf{Z}_k = \mathbf{z}_k - \mathbf{h}(\mathbf{x}_0) = \widetilde{H}_{k-1}\delta\mathbf{x}_{k-1} + \mathbf{v}_k \end{matrix}} \quad \text{(EKF)}. \qquad \text{(C.9)}$$

where $\mathbf{Z}_k = \mathbf{z}_k - \mathbf{h}_k(\mathbf{x}_0)$ is the actual measurement vector minus our best estimate of what the measurements should be based on the linearization point from the last update. A tilde is added to the measurement matrix $\widetilde{H}_{k-1}$ to indicate that it is a linearized estimate of the actual relationship between the state vector and the measurements.

After updating the EKF, the best estimate of the current state vector is updated. Exactly how this is done depends on the states being tracked, but we can write notionally that

$$\mathbf{x}_k = \mathbf{x}_{k-1} + \delta\mathbf{x}_k.$$

---

[5]The state vector might be just the target position, but it could include velocity and acceleration.

Table C.2 Extended Kalman filter update equations

EKF equation number	EKF equations (In order of update)
1	$K_k = P_k^- H_k^T \left(H_k P_k^- H_k^T + R_k \right)^{-1}$
2	$\delta \widetilde{\mathbf{x}}_{k-1} = K_k \mathbf{Z}_{k-1}$
3a	$P_k = (I - K_k H_k) P_k^-$
3b	$P_k = \left(P_k + P_k^T \right)/2$
4a	$\delta \widetilde{\mathbf{x}}_k = \Phi_k \delta \widetilde{\mathbf{x}}_{k-1}$
4b	$\mathbf{x}_k = \mathbf{x}_{k-1} + \delta \widetilde{\mathbf{x}}_k + \mathbf{d}_k$
5	$P_{k+1}^- = \Phi_k P_k \Phi_k^T + Q_k$

A key point in the EKF is that once this correction is made, we assume that the Kalman filter has done its best in producing the current correction, and the error $\delta \mathbf{x}_k$ is set to zero for the filter update at the next measurement time. That is, the state vector is zeroed for the next update. As such, the five EKF equations are enumerated in Table C.2.

There is a mixture of k and $k-1$ terms in these equations because of the linearization point being from the last update. In particular, even though the current measurements are used, the predicted measurements are based on the last update, and we assign the error to the $k-1$ index in the second equation and then update it to the current index using the transition matrix before updating the state vector. The update of the state vector is done outside of the EKF.

The "dynamics term" in step 4b is there to account for nonlinear dynamics between updates. In practice when tracking the position and velocity of a target, this step would be, for example,

$$\mathbf{v}_k = \mathbf{v}_{k-1} + \delta \mathbf{v}_k$$
$$\mathbf{p}_k = \mathbf{p}_{k-1} + \mathbf{v}_k \Delta T + \delta \mathbf{p}_k,$$

where $\delta \widetilde{\mathbf{v}}_k$ and $\delta \widetilde{\mathbf{p}}_k$ are the velocity and position components of $\delta \widetilde{\mathbf{x}}_k$, respectively, computed in step 4a.

A word of caution to the buyers of EKFs. If the point of linearization becomes too much in error, the *HOTs* are no longer small. The result is that the filter may lose lock on the truth, and quickly diverge, at which point the filter should be terminated and restarted. An example of such behavior is shown in Chap. 9. But if the initial guess at the truth is too much in error, i.e., it is out of the EKF's pull-in range, the filter may never lock onto the truth. When building EKFs, it is a worthwhile exercise to determine the pull-in range by starting the filter with more and more error in the first linearization point. The pull-in range may indeed be large,[6] but the EKF may also be quite fragile to errors in the first point of linearization depending on the degree on nonlinearity of the measurement equation and the dynamics. So one must be careful to obtain a first point of linearization that is not too far from the truth.

---

[6] As evidenced by GPS sets that initially can be started anywhere in North America, for example, and still lock onto the satellites in view and produce an accurate first fix.

There are at least two ways to determine that the filter has lost lock on the target. One is that the target is not detected for a number of observation periods (when the predicted AESA pointing angle is way off the target). Or, because Kalman filters often break sharply, the change in position from update to update may indicate a target speed that is unreasonably large. But because of Swerling RCS fluctuations, it is quite possible to miss the target on a few consecutive observations while the Kalman filter can still be on track with another detection and update. There are at least three ways to handle missed detections and still keep the filter locked. One is to note the time since the last detection, and change the update period dependent filter quantities (primarily the transition and process noise matrices) to account for the time of the missed updates. Another is to update the filter anyway on misses, but zeroing the measurements so that the state vector does not update with bad measurements. A third and similar method, and the one used in Chap. 9 simulations, is to zero the Kalman filter gain with misses.

Weighted Least-Squares Solution

Sometimes one does not want the solution to contain so much history as there is in the LKF and EKF algorithms. In this case, one can truncate the EKF, essentially computing the Covariance matrix from scratch each update. Letting the initial guess at the covariance be P_0 (for example, a diagonal matrix with 10^4 [m^2] for the position components and 10^2 [(m/sec)^2] for the velocity components), Kalman gain is computed as before:

$$K_k = P_0 H_k^T \left[H_k P_0 H_k^T + R_k \right]^{-1}.$$

Also as done in the EKF, we assume that the updates last time were perfect, setting the state vector to zero afterwards, so the state update is

$$\delta \tilde{\mathbf{x}}_k = K_k \mathbf{Z}_k = K_k [\mathbf{z}_k - \mathbf{h}(\mathbf{x}_0)].$$

The covariance matrix can also be updated to give more realistic error bounds:

$$P_1 = [I - K_k H_k] P_0.$$

On the next update these equations are repeated, starting the covariance at P_0, computing the Kalman gain, the state update, and the new covariance estimate, as summarized in Table C.3.

Table C.3 Weighted least-squares solution update equations

WLS equation number	WLS equations (In order of update)
1	$K_k = P_0 H_k^T \left(H_k P_0 H_k^T + R_k \right)^{-1}$
2	$\delta \tilde{\mathbf{x}}_k = K_k \mathbf{Z}_k$
3	$P_1 = (I - K_k H_k) P_0$

References

Analog Devices. HMC635 GaAs PHEMT MMIC driver amplifier, 18–40 GHz. Analog Devices Data Sheet. 2021;

Bayliss ET. Design of monopulse antenna difference patterns with low sidelobes. Bell Syst Tech J. 1968;47(5):623–50.

Blackman S, Popoli R. Design and analysis of modern tracking systems. Boston: Artech House; 1999.

Brown RG, Hwang PYC. Introduction to random signals and applied Kalman filtering. 3rd ed. New York: John Wiley & Sons; 1997.

Dana RA. Electronically scanned arrays (ESAs) and K-space gain formulation. Cham, Switzerland: Springer Nature; 2019.

Dana RA, Moraitis D. Probability of detecting a swerling I target on two correlated observations. IEEE Trans Aero Electronic Syst. 1981;AES–17:727–30.

ITU. Specific attenuation model for rain for use in prediction models. Recommendation ITU-R P.838-1, International Telecommunication Union; 1999.

ITU. Approximate estimation of gaseous attenuation in the frequency range 1–350 GHz. Recommendation ITU-R P.676-7, International Telecommunication Union; 2007.

Mailloux RJ. Phased array handbook. 2nd ed. Boston: Artech House; 2005.

Marcum JL. A statistical theory of target detection by pulse radar: mathematical appendix, RM-753. Santa Monica, CA: RAND Corporation; 1948.

NOAA. U.S. Standard Atmosphere, 1976, NOAA-S/T 76-1562. National Oceanic and Atmospheric Administration; 1976.

Papoulis A, Pillai SU. Probability, random variables and stochastic processes. 4th ed. Boston: McGraw-Hill; 2002.

Sherman SM, Barton DK. Monopulse principles and techniques. 2nd ed. Boston: Artech House; 2011.

Singer RA. Estimating optimal tracking filter performance for manned maneuvering targets. IEEE Trans Aerospace Electronic Systems. 1970;AES-5:473–83.

Skolnik MI. Introduction to radar systems. 1st ed. New York: McGraw-Hill; 1962.

Skolnik MI. Introduction to radar systems. 3rd ed. Boston: McGraw-Hill; 2001.

Swerling P. Probability of detection for fluctuating targets. IRE Trans Information Theory. 1960;IT-6:269–308.

Taylor TT. Design of line source antenna for narrow beamwidth and low sidelobes. IRE Trans Antennas Propagation. 1955;3:16–28.

Index

Printed in the United States
by Baker & Taylor Publisher Services